SAFE Is Not An Option

SAFE Is Not An Option

How A Futile Obsession With Getting Everyone Back Alive Is Killing Our Expansion Into Space

Rand Simberg

Interglobal Media LLC
Jackson, Wyoming

For information, contact Interglobal Media LLC at:
media@interglobal.org

This book is available at special discounts for bulk purchases from
corporations, institutions and other organizations. For information
on this, contact: bulk-purchases@interglobal.org.

Book Design, Illustration, Editing, Indexing
Cover Design and Author's Photo
By Bill Simon

Title and page heading text set in Myriad Pro — Body text set in Charis SIL
Caption text set in Arial — Footnotes and Index set in Times New Roman

ISBN 978-0-9891355-1-1

First Edition: October 2013
Revision A: January 2014

Printed in the USA

Praise for *SAFE Is Not An Option*

"Thoughtful, comprehensive, yet iconoclastic – *Safe Is Not An Option* succinctly addresses the unrealistically skewed risk-reward perception in the civilian government space sector. Speaking as a military aviator and astronaut, I believe this work wisely highlights the shortfalls of the NASA management not-truly-operational culture and offers a productive and realistic alternate viewpoint for anyone pursuing the "ad astra" dream."

Rick Searfoss
*Colonel, USAF Retired, Astronaut/Space Shuttle Commander,
XCOR Aerospace Chief Test Pilot*

——

"In 2008 I had the great privilege to fly privately to the International Space Station, where I lived for twelve days. Having grown up with a NASA astronaut father who flew on Skylab and the Shuttle, I have had a lifetime opportunity to see how flight safety has evolved in the United States. Having trained in and flown aboard the Soyuz, I have also seen the Russian approach to this same important issue. The two hardest items to control in space exploration are cost and safety. Interestingly, the Russian approach has often created both improved safety and cost in comparison with domestic strategies. If we as a global people are going to push the boundaries of humanity further into the cosmos, we must decide how much risk we should accept and how should we manage to that level of risk. Today, we seem so risk averse that we encumber the already difficult problem of space exploration with red tape, that slows down the activities, requires spiraling budgets and arguably does not improve safety in relative measure."

Richard Garriott
Private Astronaut and Computer Game Pioneer

"Since the end of Apollo, U.S. space operations have ostensibly emphasized safety first. Rand Simberg persuasively explains why that has been a mistake, and how we must change if we are to succeed."

Glenn Reynolds aka "Instapundit"
Beauchamp Brogan Distinguished Professor of Law, University of Tennessee, and co-author Outer Space: Problems of Law & Policy

————

"A pioneer at the dawn of aviation observed: 'If you are looking for perfect safety, you will do well to sit on a fence and watch the birds; but if you really wish to learn, you must mount a machine and become acquainted with its tricks by actual trial.' And since ancient times it has been known that staying in the harbor is safer than venturing onto the high seas. But nature demands tolerance of risk if the human race is to expand beyond our world. Opening the space frontier to humanity will require no less acceptance than these historical precedents, as Rand Simberg ably illustrates."

Gary C. Hudson
President, Space Studies Institute

————

"*Safe Is Not An Option* makes a strong case for changing minds and policies about our risk-averse western society and approach to future commercial space exploration. It's an excellent read. Rand tears the sheet off the elephant in the room and exposes us to the conversation in which western society must engage to remain relevant in the new millennium. The topic deserves discussion. This book is a valuable first step."

Stuart O. Witt
CEO, Mojave Air & Space Port

————

"Rand Simberg presents an intriguing case that the safety culture within the government space program, while well intentioned, is in fact detrimental to the progress of space exploration and development. The no-holds-barred approach of his viewpoint is sure to disturb the status quo, but regardless is a captivating read. Whether you agree with his viewpoint or not, *Safe is Not An Option* provides a necessary perspective for those involved in or connected to the space community."

Michael J. Listner, Esquire
Principal, Space Law & Policy Solutions, President & CEO (Interim)
International Space Safety Foundation

———

"NASA's approach to manned space flight has created the impression that such travel is inherently extremely costly. In this new book, Rand Simberg makes a persuasive case that NASA's unprecedented risk aversion is the cause of that high cost— and that such risk aversion is contrary to the history not only of aviation but of all transportation and the exploration of new frontiers. This has profound implications for the development of commercial transportation in space. Simberg offers an alternative approach which could lead space transportation to develop into an industry along the lines of aviation, rather than remaining a tiny, costly government monopoly."

Robert Poole
Director of Transportation Policy, Reason Foundation

———

"Mr. Simberg makes the compelling case that great deeds and great rewards require great risks, but NASA and my colleagues in Congress have become so risk averse in the arena of human spaceflight that we are incapable of accomplishing great deeds. America must have the stomach to let explorers and settlers willfully take on the kinds of risk necessary for opening the frontier of space to settlement under the rule of law. If we continue to overvalue that risk, or prohibit those who would willfully undertake it, then other nations with no respect for human life will be more than happy to fill that void. Left unchecked, the well-meaning, but misguided, group that promotes "safety at all cost" will continue to establish hard ceilings that we can't break through, require the expense of immense amounts of time and money, and will ultimately cost us our preeminence in space. We must not cede the high ground of space to those who do not believe in freedom. And we must respect the freedom of those individuals who are willing to put it all on the line to head over that next hill – even when that hill is in space. Mr. Simberg's book *Safe Is Not An Option* handles this sensitive issue with skill, grace, and tremendous insight."

Rep. Dana Rohrabacher
Vice Chairman of the House Committee on Science, Space, and Technology; and former Chairman of the Space Subcommittee

———

"My first thought upon reading a draft of Rand's book was, 'It's about time that someone used common sense when addressing the space safety subject.' The biggest difference between NASA in the 60s and the current NASA is how they deal with risks. In the 60s, NASA developed and flew seven new manned space launch systems (Redstone, Atlas, X-15, Titan, Saturn I, Saturn V and LM). All except X-15 were flown without fatalities. In the forty-two years since, only three new systems were flown, one Chinese, the Space Shuttle and SpaceShipOne. In spite of new safety policies, NASA's Shuttle proved to be the most dangerous way to fly outside the atmosphere. In spite of the evidence, NASA still insists on following the Shuttle model in developing future systems, which clearly hampers creativity (opportunities for breakthroughs) while providing no real improvements in safety. I applaud Rand for publishing his important research on the safety culture. This book will be referenced widely in the future and will provide the sanity that is needed while we move ahead with new technologies."

Burt Rutan
Aircraft and Spacecraft Developer

Contents

For Patricia

Foreword

I spent the weekend of August 23, 2003 aboard the International Space Station reading and digesting the official report on the destruction of another spacecraft, the Space Shuttle *Columbia*, which was to have been my ride to orbit for this flight. Instead, two weeks following the loss of *Columbia*, I was dispatched to Russia to fly on a Russian Soyuz spacecraft, to spend six months on the station while we waited to see if the Shuttle would fly again. The striking thing about the report was not the physical cause of the accident (foam loss striking the leading edge of the wing), but rather the inclusion of the lack of an overarching goal for the space program as a contributing cause. They could not have been more correct.

For years, I and other astronauts would cringe when statements were made that "safety is our number one priority." If that were really the case, then we should have given up and never flown, since sitting home on our couches would clearly be safer. The amount of risk one should take (or the amount for which safety is important), depends entirely on the rewards to be gained by undertaking the mission. Without a discussion of what we are trying to accomplish, we have no way to judge the risks we should be prepared to assume.

This book—*Safe is Not an Option*—brings this point home clearly. It not only provides the historical context of the types of risks people have assumed on other endeavors, but how this mindset of safety above all else is the end result of having no clear mission or purpose. Flying into space on current launch vehicles is not without risk, and we in the astronaut corps understood that. Obviously we were concerned with safety, which to us meant not doing something stupid. But when the risks are understood and reasonable steps have been taken to control those risks (reasonable as defined by the mission we are trying to accomplish), then it is time to light the rockets and go. That is what we were paid to do.

Those who confuse safety as being our main goal, as opposed to accomplishing the mission, are not doing us any favors.

Rand Simberg has written a book that needed to be written, and I commend him for taking on this subject. The book is not just informative and thought provoking, but also fun to read. It is aimed at the layperson, but at the same time has plenty of material for experts, and it doesn't sacrifice either technical or historical accuracy for simplicity. Policy makers, engineers, and managers who care about the development of space should read it and think about what "safety" really means. It will be well worth their while.

Ed Lu – Former Space Shuttle, International Space Station and Soyuz Astronaut/Cosmonaut and Chief Executive Officer of the B612 Foundation.

Preface

Throughout history, humans have always had to balance risk against reward. But we don't always do a good job of it, especially in modern, high-technology times, in part because statistics aren't particularly intuitive.

One of the reasons that we've made relatively little progress in human spaceflight over the past half century—despite the expenditure of hundreds of billions of dollars on it—is that we've done an absolutely terrible job of establishing an appropriate risk-reward balance. Since the end of Apollo, as a nation, we have taken an irrational approach to spaceflight-safety policy. For years the author has pointed out this irrationality, which not only misallocates resources, but keeps actual human spaceflight both very expensive and very rare.[1] Space engineer Robert Zubrin makes a similar point in a 2012 *Reason* magazine article.[2]

No frontier in history has ever been opened without risk and loss of human life and the space frontier is no different. That we spend untold billions of dollars in a futile attempt to prevent loss of life is both a barrier to opening the space frontier, and a testament to the lack of national importance of doing so. Historically, had we taken the same attitude toward safety in expanding our ecological range as we have in space, humanity would never have left Africa for Europe and Asia, colonized the Arctic, opened up the Americas, Australia, the south Pacific islands, or settled the American West. We would not have made great scientific discoveries, developed steam engines, or steam ships, or rail, or automobiles, or aircraft. In fact, we would still be sitting in the trees, gazing out at the savanna and wondering what it might have in for us.

---

1 Simberg, Rand, "Risk Aversion And NASA Don't Mix" *Popular Mechanics*, Sept. 10th, 2009 http://www.popularmechanics.com/science/space/news/4330356

2 Zubrin, Robert, "How Much Is an Astronauts' Life Worth?", *Reason* magazine, Feb. 2012, http://reason.com/archives/2012/01/26/how-much-is-an-astronauts-life-worth/1

To get some perspective and context, and recognize the absurdity of current attitudes and policies toward space activities, it is useful to look at the risk-versus-reward levels of other human activities, both historical and current, whether for exploration, science, frontier settlement, or even recreation. With an emphasis on hazards to human life, this book will provide a brief history of risk versus reward from the great Age of Exploration, to the settlement of the Americas and the development of science and transportation technology. It will then transition to a history of the early space age, and how it evolved to its current state, with case studies of Apollo, the Space Shuttle, Constellation, the International Space Station, the Commercial Crew program, and various commercial space-transportation efforts.

The book concludes with policy recommendations going forward to provide much more, and more affordable human spaceflight activities, not to just low earth orbit, but far beyond into cis-lunar space and the solar system, finally fulfilling the promise of the past half century that has always seemed to recede into the future.

A certain amount of knowledge of the history of the human spaceflight program, and particularly the history of the past decade or so, is assumed on the part of the reader in the text of the book. But for those unfamiliar with it, or who may be confused by much of the confusing or misleading reporting and commentary on it, resources are included that provide background and definitions:

- Appendix A is a sort-of glossary of the various manned American programs from 1981 to 2013, describing what things, like the Vision for Space Exploration (VSE), and Constellation, are or were, and what they aren't or weren't.

- Appendix B provides a detailed description and critique of the Space Launch System (SLS). For reasons that will become clear upon reading, it is titled, "The 'Senate' Launch System."

- Acronyms used in the text are usually defined where they first appear. For reference purposes, their definitions can be found in the "Acronym" section.

It should be noted that this book does not attempt to convince the reader that developing and settling space is important. It is assumed that readers would not have bothered to pick up, let alone crack and read the work, if they didn't already believe that. The audience of the book is not those who don't care about space, but rather those who do, or at least think they do (as evidenced by, for example, support of expensive NASA human spaceflight programs). The purpose is not to argue for the worth of such activities *per se* (that would be the subject of a different book). Rather, it is to expose the inconsistency in thinking that such endeavors are worth the expenditure of vast amounts of our national treasure, yet not also worth the risk of the loss of life.

Regardless of how much we expend to protect human life, when people fly in rockets loaded with tons of explosive propellants, and venture out into hazardous environments—with hard vacuum, high radiation levels, toxic soil, and high-velocity impacts that can easily kill in an instant—we should expect that there will be occasional loss of life, and accept it. On the menu of "out there" is discovery, adventure, awe, expanding life beyond the biosphere in which it evolved, technological advancement, the generation of wealth, and liberty. But on that menu, "safe" is not one of the options. Hence the title of this book, which shouldn't be taken to mean that safety is not desirable, or that we shouldn't do our best in our designs and operations to minimize casualties, within reason. Rather, it means that in opening up a new home for humanity, absolute safety is ultimately unavailable.

We live with the risk of injury or death in every other human endeavor, from mountain climbing to skydiving, from driving to flying. But for some reason, space-related activities are held to a different standard. Why is it that we see the death of test pilots as an unfortunate consequence of their job, but not so for astronauts? The intent of this book is to spark the needed national discussion on this topic to bring balance in terms of risk and reward and embolden us to once again move forward on the high frontier.

Prologue

It had been a successful mission, and the crew prepared to slowly start the long controlled fall home after completing the retroburn from the Orbital Maneuvering System. But unbeknownst to them, the first hint of trouble appeared a few minutes after Entry Interface, defined as 400,000 feet altitude.

The vehicle was still moving at about Mach 24 – twenty-four times the speed of sound. The yaw moment had changed almost imperceptibly, with a gentle tug to the left – there was a slight asymmetry in the aircraft's aerodynamics, but no one on the ground or in the cabin noticed it at the time.

Twelve seconds later, a temperature sensor indicated that a hydraulic brake line in the left wing was warmer than it should have been, and was slightly out of specification in that regard. As the vehicle was gradually slowing down, perceived gravity slowly growing to an indiscernible hundredth of a gee, fiery hot plasma was infiltrating through the hole in the leading edge, insinuating itself into the interior of the wing and starting to damage it, but none realized it yet.

A minute or so later, as they approached the California coast from three hundred miles out, an off-nominal rolling moment appeared, more evidence of subtle changes to the vehicle's outer shape. A few seconds later, Mission Control received a signal that several sensors were starting to indicate problems, but no one on board or in the control room was yet aware.

Half a minute later, still going Mach 23 after crossing the coast line, observers on the ground in California saw an eight-pound piece of the left wing separate from the vehicle, creating a luminescent trail in the plasma, though they didn't know at the time what they were seeing. No one in Houston or in the cabin was aware of this. About the same time, the side-slip angle exceeded any previous entry experience, as the vehicle was no longer moving in a pure forward motion. A little over half a minute later, the left elevon started to trim to compensate for the off-nominal forces, but no one noticed at the time.

A few seconds later, as the craft crossed the border into Nevada, someone in Mission Control finally noted the temperature-sensor anomalies from a couple minutes earlier.

A few more seconds after that, as the crew was doing a pressure check on their suits in preparation for landing, the brightest piece of debris was shed, but the sensors indicated nothing, and no one on board saw it.

The crew started to really sense the deceleration, about a third of a gravity, a minute or so later, as the dynamic pressure increased to forty pounds per square foot, a value that allowed the aero surfaces to take over from the small rockets that had been controlling the vehicle's attitude. Thirty seconds later, they commenced the first roll reversal to bleed off excess energy, from right-wing low to left-wing low, starting the standard series of S-turns as they approached their eventual landing site. For the veterans on board, the entry seemed to be going normally.

But a picture taken from an infrared camera in New Mexico showed some bulges in the flow field on the left wing that couldn't be explained by the vehicle attitude. No one saw it until later. The trim on the elevon was now departing sharply from a standard entry, as it fought to maintain the nominal attitude in the face of increasing asymmetric aerodynamic loads.

A little over half a minute later, alarms went off in the cabin as sensors indicated problems with pressure in the left main tire. A few seconds later, there was a false signal that the gear itself on that side was deployed and locked. After a minute or so, the crew received a call from Mission Control about the tire-pressure issue. Within a few seconds, the left elevon had lost control authority to maintain the proper pitch and roll rate. At that point, the yaw jets started to fire continuously in a vain attempt to prevent the vehicle from turning to the left. The crew could have seen the indicator lights for this, and a rapid decrease in the propellant on the gauge, but we'll never know if they did, because the signal was lost at about that time and Mission Control had received their last transmission. But the Flight Control System in the crew module started annunciating its own master alarm, as actuators started to fail, which wouldn't have gone unnoticed.

Within a few seconds, the hydraulics themselves failed, and the vehicle was out of control. It started to transition from a controlled glide to a ballistic trajectory, like a misshapen cannonball. What had been a forward-moving aircraft started to corkscrew, something that the commander and pilot couldn't have failed to observe, both from the unexpectedly changing acceleration and shifting horizon in the windshield. The crew members behind them on the flight deck probably realized that things were going south as well. A few seconds later, the cover of the left Orbital Maneuvering System pod was torn off, and the top of the cabin started to overheat.

The acceleration was starting to build up, but the crew did manage to restore the mode of the digital autopilot to auto when it was accidentally changed to manual, a few seconds later. At that point, the reaction control system was out of propellant, having fought as long as it could to keep the vehicle under control. Seeing the loss of hydraulic pressure from the failed actuators, the flight crew attempted to restart the Auxiliary Power Units that drove the pumps, to no avail. It was likely their last conscious act to save themselves.

The left wing was tearing off, shearing the softened and melting aluminum, and the noise within the cabin would have been horrific. The vehicle started to pitch up.

Within another ten or fifteen seconds, attacked by the dynamic pressure from a direction never planned by the vehicle's designers, the payload-bay doors blew off, exposing the payload bay itself to the brutal force. The hurricane of plasma separated the forebody of the vehicle, with the crew cabin, from the rest.

With all power lost, the cabin went dark other than sunlight and earthlight from the windows, but not silent, and started tumbling. The hull itself was quickly breached, and rapidly started to lose pressure. The crew had never closed the helmets of their flight suits, but it probably didn't really matter. From either the rapid accelerations or the decompression, unconsciousness likely came very quickly, which is merciful, because the heat started to melt the structure itself, spattering molten metal within.

The fiery cabin, now briefly their airborne coffin, continued its long, thirty-four mile uncontrolled elliptical fall over eastern Texas and Louisiana, with the charred remains of its inhabitants, until it came completely apart from the increasing atmospheric pressure forces and all took their own path to the earth. Ironically, and perhaps almost poetically, had it occurred an orbit later (as they had considered earlier that day for weather reasons at the planned landing site in Florida), Johnson Space Center would have been ground zero. The debris would have rained at supersonic speeds over the southeast Houston neighborhoods in which they had lived, and in which their coworkers continued to live. It might have even hit the Mission Control Center itself, where so many of the fateful, and ultimately fatal, decisions had been made.

We will never know what thoughts went through the minds of *Columbia's* doomed crew in that last minute in which they were becoming aware of their imminent peril. What we can know is that they died doing something that they had worked years to be able to do, they loved it, and thought it important. If we now fear to open up the high frontier because of what happened to them, they will have died in vain. The Columbia Accident Investigation Board (CAIB) blamed a flawed safety culture at NASA for the loss of the vehicle and death of the crew. But in a sense, as I hope to convincingly show in this book, they also could be said to have died from risk aversion itself.

Columbia's STS-107 crew: From left to right are Mission Specialist David Brown, Commander Rick Husband, Mission Specialist Laurel Clark, Mission Specialist Kalpana Chawla, Mission Specialist Michael Anderson, Pilot William McCool and Israeli Payload Specialist Ilan Ramon   —Photo by NASA 1/1/2002

Chapter 1

The Hazards Of
Exploration And Settlement

Had the early European explorers of the New World taken the attitude toward risk that contemporary NASA, American culture and Congress does with regard to space, they'd have never left the harbor.

Geographical Exploration

Half a millennium ago, even ignoring hostilities, there were a vast number of ways to die at sea, and many found most of them. Navigation was primitive, with no reliable way to know longitude, which could result in getting far off course with inadequate supplies, and no way of restocking. There was nothing resembling weather forecasting, other than the old "red sky in morning, sailors take warning" doggerel, and one could get caught by a hurricane with no way of getting away from it. Or the opposite, stuck in "the doldrums" with no wind to fill the sails and move the ship, and no fresh water. Uncharted waters near shore could contain hidden and deadly hazards on which ships could run aground, and so-called "rogue" waves (a sudden unexpected wave, much larger than the average sea state in which it appears) remain a hazard to ships to this very day. Nutrition was inadequate, with scurvy common on a diet of biscuits and dried cod, until the British later discovered that citrus would provide the missing vitamin C needed to prevent it.[3]

Knowing the hazards, Ferdinand Magellan's famous first circumnavigation expedition used five ships for redundancy, and still almost failed completely. Starting with 270 total crew,

---

3 Hence the nickname for that nation's not just sailors, but inhabitants, as "Limeys."

Ferdinand Magellan

some were killed in what is today Brazil in a failed mutiny, then a ship was lost to a sudden storm in the south Atlantic, though the crew survived. While the expedition was making its way through what is now called the Straits of Magellan to the Pacific, another ship abandoned the mission and headed back to Spain. The remaining three ships barely reached the Philippines, where Magellan himself and many others were killed in a battle with the natives[4], to the point that there were too few crew for the remaining three ships, so one of them was scuttled to allow the remaining two to continue on to Europe across the Indian Ocean. But one of the two had a leak, and stayed behind to try to repair it. When it did set sail, the crew decided to head back across the Pacific, but it was captured by the Portuguese. The lone ship that completed the circle around the world limped into Seville with a skeleton crew of 18, of which only four were from its original crew of fifty-five. Of the total original 270 to depart from Spain, 232 of them, over 85%, died on the voyage.

Even in more modern times, exploration often entailed great risk, hardship and death. Though things had advanced considerably by the mid-nineteenth century, with canned goods and ship-board clocks for navigation, exploring was still hazardous. The Franklin expedition to look for the Northwest Passage in 1845, the best-equipped polar expedition up to that time, had no survivors, and apparently

Ernest Shackleton

---

4 Interestingly, while Magellan didn't make it all the way around, the first person to actually circumnavigate the globe was his Filipino navigator, who had started from his home, headed west to Europe, and completed the journey upon his arrival in the Philippines from the east.

resorted to cannibalism at the end.[5] Even in 1914, the renowned Antarctic explorer Ernest Shackleton ran a now-famous (though possibly apocryphal[6]) ad in the London Times:

And that last was the point. Despite the dangers, there were rewards for exploring that, for the explorers, made it worth the risk. Christopher Columbus, Magellan and others were seeking routes to riches. Others sought fame and glory, or the opportunity to proselytize their religion and convert the heathens ("God, Gold and Glory"). And for some, the adventure itself was reward enough.

Advancing Science and Technology

There are other kinds of exploration than geographical, of course. Advancing science and technology is also a form of exploration, and this has also long been a risky activity for many.

It was perhaps foolhardy of Benjamin Franklin to fly a kite in a lightning storm[7], but he succeeded in making the connection between that phenomenon and electricity. Professor Georg

---

5 It is postulated that lead poisoning from the primitive canning techniques may have contributed to the expedition's failure, by impairing the crew's judgment.

6 If actually published in a London broadsheet, "honor" would be spelled "honour."

7 There remains some uncertainty of the exact protocol of the experiment that Franklin performed, and if it was as dangerous as the popular account, primarily described by Joseph Priestley based on conversations with Franklin, in which he held a key attached to the kite string in the storm, has it. But just flying a kite in a thunderstorm, even if not in direct contact with the string, is hazardous duty.

Richmann of St. Petersburg, Russia, was not so lucky. In an attempt to replicate Franklin's experiment, he was struck, blown out of his clothing, and died of electrocution – the first person to be killed playing with electricity, but certainly not the last.[8]

Pierre and Marie Curie suffered from radiation poisoning in pioneering the discovery, classification and use of radioactive elements. She died prematurely of aplastic anemia, almost certainly brought on by years of radiation. The scientists and engineers who worked on the Manhattan Project, with its nuclear pile under the University of Chicago, the world's first, took similar risks.

Jesse William Lazear

Biomedical researchers have also hazarded their lives to advance science. Jesse William Lazear worked for U.S. Army Dr. Walter Reed (for whom the medical center in the nation's capital was later named) in Cuba in 1900. A couple decades earlier, Dr. Carlos Finlay had come up with a theory that the transmission of yellow fever was spread by mosquito bites. To prove it, Lazear (secretly) allowed himself to be bitten by an Aedes aegypti mosquito. In late September of 1900 he died from the disease, and Reed later named a research camp after him.

More specific to the subject of this book, U.S. Air Force pilot Joseph Kittinger risked his life multiple times in the interest of understanding extreme flight environments and to prove out high-altitude escape systems for aircraft like the SR-71 and U-2, or the X-15. He made repeated jumps from a high-altitude balloon, to test pressure suits and parachutes. In October of 1959, on his first high-altitude jump from around 76,000 feet, the drogue chute that was to stabilize him in the lower atmosphere deployed early and wrapped itself around his neck, putting him into a spin that caused him to lose consciousness. As he passed out he had the expectation he would not survive, but after his emergency chute

---

8 In modern times, such an event is often preceded by the immortal last words, "Hold my beer" and "Watch this."

opened automatically at about 10,000 feet, he came to and landed safely. His next test, from about 75,000 feet, went more smoothly, and they moved on to the ultimate jump from over 100,000 feet in August of 1960 at White Sands Missile Range in New Mexico.

Preparation started days before with a special diet to minimize intestinal gas that, at low pressure, could inflate him like a balloon. Hours before flight, he began pre-breathing pure oxygen to eliminate nitrogen in his blood that could cause it to boil (the condition that divers call "the bends") if he lost pressure in his suit. He then suited up and started the ascent. At around 43,000 feet he noticed that his right glove wasn't pressurizing. He knew that as a result his right hand was going to painfully swell as he ascended, but decided to continue on

Col. Joseph Kittinger II

because he would still be able to control the vehicle with his left. As he ascended through the troposphere, he passed through the region of greatest danger of his fragile balloon popping by becoming brittle from the cold temperatures. At 60,000 feet he started to exceed the balloon's drag limit in ascent speed, and had to release some helium to slow it down. He finally reached the balloon's maximum altitude, a little over 100,000 feet. Blinded by the unfiltered sun and with his right hand numb and swollen from lack of circulation, he waited for the unanticipated clouds below to clear. After several minutes he finally got the word from the ground that it was go for the jump, and so he did.

His stabilization canopy opened as scheduled, but he momentarily felt a choking sensation, and feared a repeat of his first balloon jump. But it eased, and he continued to free fall, setting a time record that stands to this day.[9] His main chute

---

9 In 2012, daredevil Felix Baumgartner beat both his altitude and speed record, going supersonic for the first time. But the freefall record still stands, because Kittinger's fall was slower and spent more time before opening his chute.

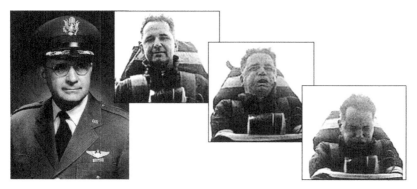

Colonel John Paul Stapp in high-speed rocket-sled acceleration and deceleration tests

opened at about 20,000 feet, and the worst was over. The total time of descent was over thirteen minutes. He had a hard landing that bruised his leg, but the swelling in his hand went down after a couple hours, and he set an altitude and speed record that stood for over half a century.

But while Kittinger was a great research pioneer, probably no single individual has advanced the safety of aerospace (and other) vehicles than John Stapp, and no one ever abused their body more in the service of that goal. A medical doctor with a PhD in human biophysiology, Colonel Stapp, first with the Army Air Corps and later with the Air Force as a flight surgeon, personally plumbed the limits of physical endurance in acceleration, deceleration, depressurization, and high winds, to determine how to maximize the probability of surviving adverse incidents in combat aircraft, spaceflight, and even auto accidents. He subjected himself to up to 45 gravities of forward acceleration ("eyeballs in") and lesser, but still unimaginable deceleration ("eyeballs out") on a rocket sled. Twice, in two separate sled runs, he broke his right wrist. He flew aircraft almost six hundred miles per hour without a canopy to see if a pilot could survive the wind blast of an ejection at high velocities. He subjected himself to rapid decompression in a hypobaric chamber to determine the degree to which a human would suffer from the bends in the event of a high-altitude (or vacuum) emergency. As a result, we have a much better understanding of requirements for seat design, restraint harnesses and other parameters that have greatly increased the safety and survivability of all manner of high-speed transportation. But he paid a heavy price. He was a tough old bird—he died in his bed at the ripe old age of eighty nine—but he suffered throughout his career and life from multiple fractures of bones in his limbs, detached retinas, and burst blood vessels in his eyes, resulting in partial permanent blindness.

The Hazards of Settlement

After first opening up new territories by exploration, settling them was often no less hazardous. Let's just consider North America alone.

The first English colony, on Roanoke Island in what is now the state of North Carolina, was established with 116 settlers in 1587. Its founder returned to England to get more supplies, but didn't get back for three years. When he did, the colony had completely disappeared. It is thought that a couple dozen of them left for Hatteras Island, a few miles to the south, to live with the Siberian-Americans[10], and recent findings indicate that the rest may have moved inland, but no one knows to this day, over four hundred years later.

The next attempt was Jamestown, a decade and a half later, on an island in the mouth of the James River in southern Virginia (near modern-day Newport News and Williamsburg). It started in mid-May, 1607, with 104 settlers, all men and boys (not an auspicious way to begin a "settlement"). Half of them were "gentlemen." That is, they had no useful skills on a frontier in terms of building, clearing, farming or other physical labor. However, they were skilled at fighting, and that in fact turned out to be a useful skill indeed. Between Indian attacks and diseases arising from living in a mosquito-infested swamp in the southern summer, within eight months their numbers had dwindled to 38, losing more than half of the colony. Over the next decades, there was continuing death and attrition from famine, pestilence and warfare with the

---

10 There are no humans native to the Americas—we all came from somewhere. Henceforth they will be referred to as Indians.

Indians, but the foothold was established, and Virginia became the first English colony.

Most going to Virginia were seeking wealth (initially from tobacco), but in 1620 one group headed out from the east midlands of England seeking instead religious freedom. The ship went off course, and ended up much further north, in what is now Massachusetts, on Cape Cod. Landing in late November, they had difficulty building shelter, and 45 of the original 102 died in the first winter.

Once the early footholds were established, though, the eastern seaboard became a much safer place to live. But once again, settlers started moving west, and once again, they had high attrition rates. For example, in the settlement of Kentucky, thousands died from accidents, illness and Indian attacks (the latter didn't end until the American victory at Fallen Timbers in 1794). Whole families were wiped out, and often villages were decimated, by disease.

It continued as the settlers headed across the Mississippi into the plains and deserts of the west. Some sought gold, some sought land, and others, like the Pilgrims who initially settled Massachusetts, sought religious freedom. For example, the Mormons headed to the frontier to be allowed to practice their homegrown beliefs without hindrance. All groups were beset by hostile Indians, there were accidents galore, and raging epidemics of typhus and cholera. In one particularly gruesome and infamous incident in which they did almost everything wrong, a pioneer party (the Donners) was trapped in the high Sierra in an early winter storm and some resorted to cannibalism to survive.

Back east, about the same time, another emigration had huge casualty rates as the Irish left their home country during what they called The Hunger. They weren't seeking freedom, or riches, but simply food, and could barely afford passage even when crammed into what became known as "coffin ships":

> their passengers "were only flying from one form of death." While they may have left starvation behind, many of these passengers were already in extremely bad health after a year

or more of inadequate nutrition and exposure to illness. With their physical state already desperate, the last thing they needed was to be crammed into overcrowded, insanitary conditions with hundreds of others.

Just one case of typhus, which was rampant among the poor at this time, could spread like wildfire in the conditions on the coffin ships, and many were to die before, or shortly after, reaching the other side of the Atlantic. Others drowned when their ships were overwhelmed by ocean storms or fell upon rocks.

The ships that survived the Atlantic crossing arrived at the quarantine station of Grosse Isle, the Canadian immigration point and depot set up in the Gulf of St. Lawrence (Ontario) in 1832, to contain diseased immigrants to British North America. Statistics for just one month - July 1847 - indicate the horrors that were being endured. Ten vessels arrived that month; of the 4,427 Irish immigrants that had started their journeys (all had departed from either Cork or Liverpool), 804 had died on the passage while 847 were sick on arrival.

By the end of 1847, the awful toll could be calculated from the 200 immigrations ships that had made the crossing. Of 98,105 passengers (of whom 60,000 were Irish), 5293 died at sea, 8072 died at Grosse Isle and Quebec, 7,000 in and above Montreal. In total, then, at least 20,365 people perished (the numbers of those that died further along in their journey from illnesses contracted on the coffin ships cannot be ascertained)—one-third of each vessel's passenger list.[11]

All of these people had their reasons to emigrate, and many in the future may have similar reasons to emigrate to space as the technology allows them to, and conditions evolve on the home planet. The American frontier had air, water, soil, food on the

---

11 "Coffin ships: death and pestilence on the Atlantic," *Irish Genealogy*, http://www.irish-genealogy-toolkit.com/coffin-ships.html

hoof, and other resources that allowed its settlement. So did the Arctic, but it required a more advanced technology level (fire, furs, spears, fishhooks) to be conquered by naked humans originally from the tropics. Similarly, space has resources, but they require advanced technology to utilize, and we'll have to manufacture our own air, water and soil (assuming that we grow food that way). It is a much harsher frontier than anything our ancestors have faced, and it's unreasonable to think that it will be opened up without loss of human life, despite the lack (as far as we know) of hostiles. To delude ourselves that it can be is to keep it forever closed to us.

The harsh environments of space, the moon and Mars

Chapter 2

The Death Toll
From Transportation

The history of the development of modern transport itself, that will be necessary for the opening of the space frontier, is a bloody one.

Early Powered Transport

The first vehicle to be moved by something other than animals (including humans), or wind or current, was the steamboat, in the late 18th century. Steam engines first propelled boats, then it was applied to rail locomotives early in the 19th century. In both cases, because we didn't at that time understand good design practices, the limits of materials, or the pressures involved in creating and controlling steam, there were many mishaps. And by "mishaps," I mean events that caused many hundreds of passengers and crew on steam-powered vehicles to meet their maker. Boiler explosions were a common occurrence throughout the nineteenth century in both England and America. Hewison notes that there were 122 of them on locomotives alone in the UK.[1] As Samuel Clemens (aka "Mark Twain") described in his classic semi-autobiographical book, *Life on the Mississippi*, steamboat explosions, both in the boilers and fireboxes, were all too common on the rivers and lakes of the young American

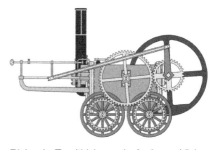

Richard Trevithick and Andrew Vivian steam locomotive 1802

1 Hewison, Christian H. (1983). Locomotive Boiler Explosions. David and Charles. ISBN 0-7153-8305-1.

nation. In 1865, at the end of the Civil War, after a faulty
repair, one of the *Sultana's* four boilers exploded near Memphis,
Tennessee. The ship went down in "ol' man" river, killing many
hundreds of passengers (mostly returning former Union prisoners
of war), either from the violence of the explosion itself, from
burning or drowning if trapped, or from drowning or hypothermia
in the still-frigid flooding river waters (many, perhaps most, were
unable to swim). The ship had only been designed for less than
four hundred passengers. It was the greatest single maritime loss
of life in the nation's history up to that time. But we didn't end
steamships, or shut down the railroads, because they were too
valuable, too important. Most judged the risk worth the payoff,
and those who didn't led poorer, if safer lives.

As design, materials and maintenance improved, the rate of
boiler and firebox incidents was reduced. The numbers dropped
dramatically in the twentieth century with the advent of the
American Society of Mechanical Engineers (ASME) code for
the design of boilers and pressure vessels in 1915, spurred by a
spectacular explosion at a shoe factory in Massachusetts in 1905
that leveled the building and killed dozens. In fact, steam engines
were considered to have become safe enough to deploy them as
motors for the infant automobile industry. Steam-propelled cars
were particularly popular with women, who didn't have to crank
an internal-combustion engine to start them (beyond their strength
in many cases, and a hazardous activity in itself—many an arm by
men and women alike was broken from kickback). The Stanley
Steamer was the most noted example, but for the most part, steam
cars didn't survive the invention of the electric starter, at which
point internal combustion dominated to the present day, even
with the burgeoning advent of hybrids and electric. And despite
the fact that millions of people have been killed by cars, however
engined, over the decades, they continue to drive them. But let's
move on to the progenitor of the space industry—aviation.

Aviation

It should be no surprise that a form of transportation that
entails higher speeds than any previous one, generally far above

the ground, has a bloody development history. Not counting those killed in glider accidents in the nineteenth century, such as Otto Lilienthal (a major influence on the Wright brothers), the first person to be killed in a powered aircraft accident was Army Lieutenant Thomas Selfridge, with Orville

Orville piloting the Wright Flyer

Wright at the controls of the Wright Flyer III, almost half a decade after the brothers' first flight. One of the craft's two propellers came loose and tangled up the wires that adjusted the wing shape, resulting in a loss of control. Wright survived with broken bones. An Army Air Corps base in Michigan was later named after the (literally) fallen soldier. It was just the first of many deaths in the air, or rather (as was most often the case other than war) in abrupt contact with the ground.

In the following decade, not even counting all the aerial combat deaths in what was then called the Great War (later designated World War I, after we experienced a second one), a number of aviation pioneers died. England's first aviation fatality was Charles Stewart Rolls, one of the founders of Rolls Royce, when his Wright biplane broke up in the air in 1910. The same year, the first man to fly over the Alps died of injuries sustained when his Blériot crash landed in Milan, Italy. Eugene Ely, the first naval aviator and first to land a plane on a ship, died in a demonstration flight in his Wright Flyer at the Georgia State Fair in 1911. The first man to fly across the continental United States, Calbraith Perry Rogers, was also the first pilot to be killed by a bird strike when a seagull hit his Wright EX at an air show in Long Beach, California in 1912. A few months later, the famous aviatrix Harriet Quimby was thrown out of her Blériot

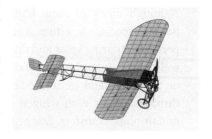

Blériot Monoplane (replica)

monoplane at an air show in Dorchester, Massachusetts. The next year, Samuel Franklin Cody and his passenger died when a floatplane of his own design broke up at five-hundred feet altitude. In 1915, Lincoln Beachley plunged into San Francisco Bay when he tore a wing off his Taube monoplane in a high-gee aerobatic maneuver.

The carnage continued into the twenties. A movie stuntman died in Los Angeles in 1920 when he was blinded by the Klieg lights and couldn't pull out of a dive. Roald Amundsen, the famous polar explorer (for whom the south pole research station is partially named), was killed with several others in a Latham 47 flying boat in 1928 while attempting a rescue of a stranded airship crew in the Arctic.

In fact, with the advent of the Curtiss JN-4 "Jennie" aircraft, in which almost every American military pilot trained, and that

The Curtiss JN-4 "Jennie"

became available after the war as surplus for as little as $200 (original cost was $5,000), the death toll accelerated. The owners of the ubiquitous aircraft, so-called "barnstormers," would fly into town and offer daring (and dangerous) aerial demonstrations, and rides to the locals for a few dollars or so. The early part of the decade was a peak era for crazy stunts the show men (and show women) would perform:

Although many of them handled all their own tricks, others became specialists, either stunt pilots or aerialists. Stunt pilots performed daring spins and dives with their planes, including the well-known loop-the-loop and barrel roll maneuvers. Aerialists, on the other hand, performed such feats as wing walking, soaring through the air with winged costumes, stunt parachuting, and midair plane transfers. Essentially barnstormers, particularly the aerialists, performed just about any feat people could dream up; there seemed to be no limit to what they could accomplish. While some played tennis, practiced target shooting, or even danced on the wings of planes, others such as Eddie Angel did their own unique stunts. Angel's specialty was the "Dive of Death," a nighttime jump from a plane that barnstorming historian Don Dwiggins describes as "a free-fall" from 5,000 feet, while holding a pair of big flashlights."[2]

Bessie Coleman, the first black aviatrix, was thrown from her Jenny while practicing a stunt for a show, and her pilot then crashed as well after a loose wrench jammed the controls. But despite the danger, many aviation pioneers got their start as barnstormers, including Charles Lindbergh, Wiley Post

Lillian Boyer doing the "breakaway"

(who was killed in the thirties in a flight that also took the life of popular entertainer Will Rogers), and Florence Lowe (aka "Pancho" Barnes, who later became famous not just for her flying feats but as the retired proprietor of her bar in Rosamond, California, after being featured in Tom Wolfe's book on the early space program, *The Right Stuff*).

But the stunts, rides, and actual transport were separate markets, and in fact the decade also saw a huge improvement in aviation safety. One of the largest of the so-called "flying circuses," run by Ivan Gates, reportedly introduced a million passengers to aviation with their "one-dollar rides," with no serious injuries between 1922 and 1928.

Despite this, however, pressure built throughout the decade for regulation of the industry for the transport market. This arose from two sources: legitimate concerns about passenger safety, and rent seeking on the part of those in the fledgling airline industry

---

2 Onkst, David H., "Barnstormers," US Centennial of Flight Commission, http://www.centennialofflight.gov/essay/Explorers_Record_Setters_and_Daredevils/barnstormers/EX12.htm

who wanted to create barriers to entry to reduce potential cheap competition from the barnstormers. The first significant legislation in this regard was the Air Commerce Act of 1926, in the Coolidge administration. It designated the Department of Commerce responsible for testing and certificating pilots, establishing and enforcing safety rules, issuing certificates of airworthiness for aircraft, and accident investigation. However, it was also to help promote the growth of the industry, a role that would be in tension for many decades with its responsibility to ensure passenger safety. As a result of the legislation, the department established an aeronautics division, which was the ancestor of the present-day Federal Aviation Administration, now in the Department of Transportation, whose dual role of promotion and safety ended after the ValuJet crash in the Florida Everglades in 1996. As of now, with regard to aviation (but not space, as we'll see), the industry is considered sufficiently mature that it can survive an FAA focused only on minimization of passenger risk.

With the new law, as well as increasing regulations of the stuntmen and stuntwomen themselves (e.g., wing walking without a parachute was outlawed, ironically increasing the risk of falling due to the bulky nature of the equipment in moving among struts and wires), the days of the barnstormers came to an end by the thirties.

In addition to the regulations, technology advances were improving safety as well. The National Advisory Committee on Aeronautics (NACA) had been formed in 1915 in response to a perceived lag by the U.S. in aviation, where it had been invented, relative to the Europeans, particularly the French (who as a result provided us with many aviation terms, such as fuselage, empennage, and nacelle). The predecessor to NASA, it supported the industry by performing basic research in areas such as best wing and propeller shapes, cowling designs, propulsion, navigational aids (Jimmy Doolittle, prior to his famed Tokyo raid in World War II, was instrumental, so to speak, in developing instrument flying), radar for crash avoidance, and other technologies that helped reduce the risk of an aircraft accident.

With this regulatory and technological infrastructure in place, from the twenties on through the end of the century, aviation passenger safety incrementally improved as a result of lessons learned over the decades from accidents. It has in fact reached the point that, in the U.S., it has now been over a decade since the last fatal crash of a major airliner. But today's regulations couldn't have been written then, because we didn't have the experience to know how to do so. Almost all of the current regulations are figuratively written in blood. Moreover, we didn't have the technology at the time to make them practical. So on that note, we come to space, the final frontier, and its peculiar and contingent history.

Chapter 3

The Initial Space Age

As Tom Wolfe entertainingly documented in his classic book *The Right Stuff*, if we hadn't been engaged in a cold war against a totalitarian superpower with nuclear missiles, it seems likely that a natural path to spaceflight would have been to expand the performance envelopes of the

X-15 Rocket Plane

high-powered supersonic aircraft and rocket planes like the North American Aviation X-15 research aircraft. Both civilian and military test pilots were risking their lives heading to the top of the atmosphere in that vehicle during the late fifties and early sixties. One, NASA pilot Jack McKay, flipped his aircraft after an emergency landing on a dry lake from an engine failure, and suffered back injuries that eventually forced his retirement. Scott Crossfield, a test pilot for North American, the aircraft's manufacturer, had to make an emergency landing with a fogged windshield he had to continually clean with his sock, after one of the engines exploded shortly after release from the Boeing B-52 carrier aircraft. He had another close call when a stuck valve resulted in another engine explosion in a ground test. Air Force pilot Michael Adams was killed when his aircraft broke up from a violent pitch oscillation after recovery from an inverted flat spin.[1]

One X-15 pilot who survived the program was a former Navy aviator and Korean War veteran, a civilian employee of first the NACA then NASA as one transitioned to the other in the late fifties. A few years later, he would become the first man to walk on

---

1 "X-15 Test Pilots," NASA Fact Sheet, Langley Research Center, http://www.nasa.gov/centers/langley/news/factsheets/x-15_2006_2.html

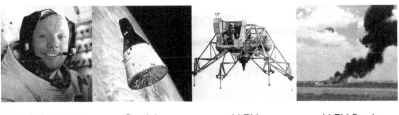

Neil A. Armstrong   Gemini   LLRV   LLRV Crash

the moon, but not before surviving a spinning Gemini spacecraft, and a failure of a lunar-lander simulator, called the Lunar Lander Research Vehicle (LLRV), by ejecting from it and parachuting down.[2] These were the sort of men from whom the initial astronaut corps was selected, and they were men quite familiar with the dangers of high-velocity, high-altitude machinery, attending funerals of their colleagues on a regular, even weekly basis.

But with the national space panic induced by Sputnik in the late fifties, the slow and steady approach to developing spaceflight technology was abandoned in an effort to retake the lead in a race in which the Soviets were perceived to be ahead. They had launched the first satellite, much larger than any the U.S. was capable of[3], the first dog, and in 1961, the first man into space, and not just into suborbit, as the first two American flights were following Gagarin's historic achievement, but all the way into orbit.[4] Because of the urgency, the nation chose to conquer space, at least initially, not gradually with faster and higher aircraft, but almost immediately by strapping men to the top of modified

---

2 In August of 2012, Neil Armstrong, the person being described, passed away at the age of eighty two as a result of complications from heart surgery he had undergone a few weeks earlier.

3 This was primarily due to the fact that use of the Atlas, whose performance was comparable to the Soviet rockets, had initially been forbidden for use as a satellite launcher for political reasons, to maintain the appearance of a civilian space program.

4 There is some controversy about this, because it was only a single revolution, and some consider it to have not made it quite all the way around, because he landed west of his launch site due to earth rotation. But his capsule did achieve orbital velocity (he had to do a retrofire to enter), and did get more than all the way around in an inertial reference frame. Also, he didn't land in the capsule, but jumped and parachuted from it to land, but this takes nothing away from the feat.

Mercury/Redstone   Mercury/Atlas   Gemini/Titan II   Dyna-Soar/Titan III
(never flew)

but existing intercontinental ballistic missiles (ICBMs), or to use another phrase, unreliable munitions.

Early ICBMs (and the intermediate range IRBMs) were not designed to be highly reliable, because to do so would have dramatically increased their costs and delayed their deployment (many hundreds of them were built), and it wasn't necessary for their mission. They were designed to be launched in massive numbers, and if a few out of a hundred didn't make it through, that was all right, because they were often redundant in their targeting (that is, more than one missile would be aimed at a key target). At the time, the reliability of the Titan I was estimated to be only 80% or so (that is, one in five would not deliver its payload to the designated target). The Titan II was estimated to be better, a little over 90%, because though eight of its initial thirty-three test launches were failures, it had a successful string of thirteen at the end of the test program. This resulted in its selection for both the Air Force's X-20 Dyna-Soar manned spacecraft (which never flew to orbit) and Gemini.

So the early manned spaceflights were performed on modified versions of missiles (specifically, the Redstone and Atlas for Mercury and Titan II for Gemini). But what was good enough for a weapon as part of a fusillade of dozens or hundreds wasn't perceived to be for a single flight carrying a human, particularly with recent memories of nationally televised ignominious failures of rockets on the launch pad. In fact, Mercury Seven astronaut Deke Slayton said later, "If you had told me that by the end of the Mercury program we'd have gotten all our manned Atlas launches off without using the escape system, I'd have said no way."[5] Thus was born the pernicious (and now obsolete) concept of "man rating," which confuses the space industry and obfuscates space-policy discussion down to this very day.

But it made sense at the time, particularly given that we were engaged in a race with the Soviets in what was perceived to be an existential war, if a "cold" one. It was a flexing of both sides' technological muscles, and needlessly killing astronauts would have been seen as an extreme technological weakness. So the missiles to be used to launch NASA's space travelers were modified, "man rated" to make them as safe as reasonable with the existing technology, while still getting their crews into orbit, and cost be damned.

The approach to do so was two pronged.

First, they would attempt to increase the reliability of the specific missile, with rigorous quality control, including documentation of every component, in some cases all the way back to the mine from which the ore of the metal from which they were constructed was originally excavated. In addition, they would add redundancy in key areas, such as guidance, to increase the probability of success. They also beefed up structure, increasing the structural margin from 1.25 (twenty-five percent) to 1.4 (forty percent), to minimize the risk of breakup during ascent.

However, it should be noted that this primarily applied to the Redstone and Atlas rockets, which had to be specifically modified after the fact. The decision to use Titan II for manned spaceflight had been made during its development, and this resulted in solving problems such as "pogo" (a longitudinal oscillation induced by coupling between the vehicle's structure, thrust and its propellant flow) that might have been acceptable for Air Force weapons delivery, but not for crews, resulting in a more reliable vehicle for all mission applications.

The second prong was to add systems to the crew vehicle that would offer its passengers an escape opportunity in the event of a

5 Slayton, Deke and Cassutt, Michael, "*Deke!: An Autobiography*" Forge, 1995, p. 84.

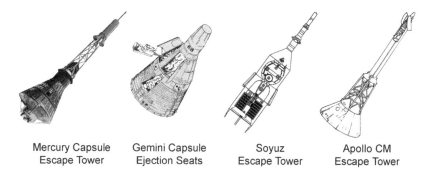

| Mercury Capsule | Gemini Capsule | Soyuz | Apollo CM |
| Escape Tower | Ejection Seats | Escape Tower | Escape Tower |

mishap, all the way from ignition until the vehicle was well out of the atmosphere. In Mercury and Apollo, this consisted of a launch tower on top of the capsule that would rapidly accelerate it away from a rocket that was out of control or exploding beneath it. For Gemini, they relied on traditional aircraft ejection seats, though much more powerful (and dangerous) to rapidly accelerate the crew away from an exploding liquid-fueled rocket. It should be noted that, as with aircraft, such measures were hardly without risk themselves. The use of an ejection seat has been properly characterized as "attempted suicide to avoid certain death." For instance, in a test of the Gemini seat, the hatch failed to blow, and the test dummy's head was slammed into the side of the spacecraft. In all cases, the rockets themselves also required further modification, to add sensors to warn if and when the rocket was starting to catastrophically fail, so that the abort sequence could be initiated in time.

In the case of the Saturn, the phrase "man rated" started to become even more fuzzy than it was for the Titan, because the original meaning was to take an "unsafe" vehicle and improve its safety, whereas the Saturn was designed from the beginning with a requirement that it be capable of carrying humans relatively safely (it was never planned to be a weapon), and its design incorporated lessons learned from earlier missile updates. But despite this, it was still considered to be man rated and, combined with the abort system on the Apollo capsule, it had all the capabilities that man rating implied, including "zero-zero" (zero altitude, zero velocity, i.e., on the pad) abort capability. Ironically, considering the danger to which the astronauts had subjected themselves as aircraft test pilots, strapping themselves to the top of a rocket was probably one of the safest things that they had ever done in their careers.

As it turned out, the launch abort systems were never required on any of the American flights in the sixties and seventies, though one was almost activated on Gemini 6, when an umbilical plug shook loose on the pad after first-stage ignition, starting up the flight computer prematurely (it was supposed to be unplugged by the vehicle actually leaving the launch pad). When it woke up and sensed no acceleration, it automatically aborted the launch and shut the engine down. Flight rules called for an ejection in a flight abort, but Wally Schirra, the flight commander, no doubt recalling the fate of the test dummy, decided to stay with the spacecraft, given that neither he or pilot Tom Stafford had felt any actual upward motion. The move (or rather, lack thereof) may well have saved their lives, and illustrated the devil's choice that must often be made in attempting to minimize risk.

Ironically, the only time that NASA astronauts died in the sixties[6] was on the launch pad in January 1967, when Gus Grissom, Ed White and Roger Chaffee asphyxiated in a ground test of their Apollo 1 capsule after a spark started a fire in the oxygen-rich atmosphere. This resulted in a change to the cabin-atmosphere

Grissom   White   Chaffee
Perished in the Apollo 1 fire in 1967

composition for future flights, though it's unlikely that the accident would have occurred in an actual flight even with the mix that was deadly at sea level, due to the reduced pressure in space that would have inhibited combustion. The accident also uncovered a number of other, unrelated problems with the capsule design that resulted in both a management and design overhaul that doubtless made the first lunar landing possible two and a half years later.

---

6 Other than in non-space-related accidents, such as flight-training crashes in aircraft. Several astronauts died this way, as did Soviet Yuri Gagarin, the first man in space.

Apollo 8 went to the moon with only the Command and Service Modules (CSM) and therefore did not have the "LEM Lifeboat" option. (Note: Image is Apollo 15)

Lovell   Anders   Borman

The crew of Apollo 8: The first men to go to the moon and orbit it

However, despite all of the precautions, NASA did demonstrate its willingness to risk the lives of its astronauts, when in a daring mission, it won the space race in December of 1968 with the Apollo 8 mission around the moon. What was daring about it?

The previous April, there had been a partial disaster during an early test of the new Saturn V rocket, whose express purpose was to send astronauts to the moon. It suffered from the same "pogo" problems that had earlier afflicted the Titan, almost shaking the vehicle apart during ascent, with some structural failure in the first stage. Two of the second-stage's five engines failed, and the single third-stage engine failed to reignite in orbit. Wernher von Braun's team went to work to sort out the problems, and a few months later, after some ground tests, declared it ready to fly again. NASA was under some pressure because there were rumors that the Soviets were going to send some cosmonauts to

Russian Zond Spacecraft

Lunar Excursion Module (LEM)

circumnavigate the moon with the Zond spacecraft by the end of the year (they had already sent some animals on such a trip).

"[A]nd from the crew of Apollo 8, we close with good night, good luck, a Merry Christmas—and God bless all of you, all of you on the good Earth."

— Frank Borman

While it wouldn't have been a loss of the space race, the goal of which was to land on the moon, and not just fly around it, being beaten to that next first would have been another blow to the national psyche after Sputnik and Gagarin, and the first space walk. The Lunar Excursion Module (LEM) wasn't ready yet, and not expected to be until the spring of 1969, so NASA decided to scrap their plan of doing an earth-orbit rehearsal, and instead decided to go for the moon on the very next flight of the Saturn V, and *without another unmanned test flight* despite the problems on the previous flight. They were willing to throw the dice, and the astronauts (Frank Borman, Jim Lovell and Bill Anders) were willing to risk their lives, because it was *important*. The whole purpose of the program was to demonstrate that our system was superior to the Soviets, and to be afraid to fly would have rendered it pointless. It is hard to imagine today's NASA taking such a risk with its astronauts' lives, because nothing NASA is doing today is perceived as being sufficiently important. This is a theme we will come back to as we get to the discussion of the modern era of spaceflight.

And in fact, the flight went perfectly, resulting in the iconic moment when the astronauts read to the world from *Genesis* on Christmas Eve, 1968 (another moment that would be hard to imagine today, given the politically correct tenor of the times) as they circled our sister world from a quarter of a million miles and a light second away. It also brought back an equally iconic photo by Bill Anders of our seemingly fragile beautiful blue home planet, a little bubble of life in the sterile blackness of space, that later adorned dorm rooms across the nation and resulted in the first Earth Day not long after.

NASA's gamble paid off, in a sense. That was the moment that the space race was essentially won. The Soviets canceled their own manned lunar flyby plans, and pretended that they'd never been racing, though they continued to secretly plan landings into the early seventies. Unfortunately, that was also the moment that space became unimportant, or at least enough so not to take bold risks any more. But it continued on inertia, and (at least partly) in honor of assassinated President John Kennedy's memory. So much had been spent that it would have made no sense, once the hardware had been proven, to not actually send men to actually land on the moon, and the nation did so eight months later.

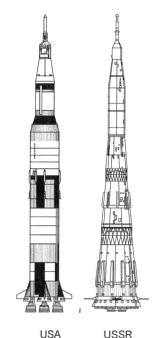

USA   USSR
Saturn V   N1
Moon rockets
Note 6' man (at bottom) for scale

On Apollo 12, the flight after the historic one that first landed men on the moon, the crew got a scare when the Saturn was struck by lightning about half a minute after lift off. The Saturn continued to function, but the strike caused a loss of power in the Command Module and, a few seconds later, another strike knocked out the navigation system. Fortunately, within minutes John Aaron, the Electrical and Environmental Systems (EECOM) member of the Mission Control team, came up with a solution that got them back on line and the flight continued successfully. This incident resulted in new launch constraints for weather, with new rules dictating a launch scrub if lightning is in the area during the launch window. (Many Space Shuttle launches were delayed for this reason).

All this time, of course, the Soviets were rumored to be having their own safety problems (rumors that were confirmed in the 1990s, after much more information became available in the wake of the fall of the Soviet Union). For instance, we now know that dozens of rocket technicians were killed, including the program manager, and many more injured in a horrific explosion of burning acid at Baikonur in the early sixties, when a huge fueled hypergolic ballistic missile prematurely started its second stage while people were still working on it on the pad.

A chute failure killed a cosmonaut in 1967, and in 1971, they lost a crew of three on entry when a relief valve to the cabin stuck open after separation from their service module and evacuated it. None of them were wearing space suits, and when the capsule came down they were found dead of decompression (these are technically the only space travelers to have actually died in space to date, if one considers *Columbia* to have not been in space when it broke up on entry in 2003). As a result, crews were reduced to two for that vehicle to allow the cosmonauts to wear suits, and later versions downsized interior equipment to allow three to be suited during flight.

The Soviets also had the first (and so far, last) in-flight abort in 1975, when on a planned trip to the Salyut 4 space station, the third stage of their Soyuz rocket failed to properly separate from the second stage, but ignited regardless. The trajectory they were on when the capsule finally separated resulted in a 20 + gee entry, and they landed on a mountainside and tumbled down it, the fall interrupted only when the parachute snagged on some bushes just before they would have plunged over a precipice. Vasili Lazarev suffered career-ending internal injuries, but his crewmate Oleg Makarov went on to fly future Salyut missions.

Unlike the Americans, the Soviets once actually had to use their launch abort system, which wasn't an ejection seat, but rather an escape tower similar to the design used by Mercury and Apollo. In 1983, the Soyuz T-10-1 mission did an abort of 15-17 gees when the rocket blew up on the pad. Gennadi Strekalov and Vladimir Titov landed safely by parachute. They were shaken up, but uninjured, and went on to fly again in the following months.

The closest that NASA came to losing astronauts in flight during Apollo was not during ascent to space, but on the way to the moon on Apollo 13 (commanded by Jim Lovell, who had circled that world in Apollo 8), when a liquid-oxygen tank exploded in the Service Module, resulting in the loss of much of the life-support consumables needed to get them to the moon and back. They were saved only because they could use the resources of the LEM, by creating new procedures and modifying the hardware literally on the fly.

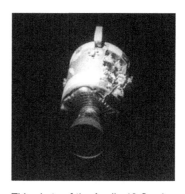

But for all the ingenuity and heroism displayed both on the ground and in space in that incident, Apollo 13 created unfortunate myths about NASA that persist to this day.

This photo of the Apollo 13 Service Module (after it was jettisoned prior to Command Module entry) shows the damage caused by the oxygen-tank explosion

First of all, the lead flight director Gene Kranz later popularized by the movie was actually a composite of the other flight directors (they actually did have shifts) and gave short shrift to Milt Windler and Gerry Griffin (Glynn Lunney was depicted in the film, but as a more minor character).

More substantively, despite the fact that it was the title of his autobiography, Kranz never uttered the words during the actual events that "failure is not an option." And that's a good thing, because while the phrase sounds inspirational in a movie in which things come out right in the end, and could bolster morale in an emergency, it is a disastrous basis for business or technical decisions. As Peter Stibrany, chief engineer of the MOST astronomy satellite once said, if failure is not an option, success can get very expensive.[7] It can also become rare, because fear of failure can often result in an unwillingness to even make the attempt. As we'll see in the coming sections, there has in fact been a lot of this at NASA since Apollo, and it largely accounts for why spaceflight remains costly and infrequent.

But most importantly, Apollo 13 established a myth about NASA: that it could accomplish anything, and always get its

---

7 Spencer, Henry, personal email, personal recollection.

astronauts back if it stepped up and displayed "the right stuff." Unfortunately, this notion doesn't take into account the actuality that, while everyone involved did everything they needed to do to save the crew, there was also a lot of luck involved. The oxygen tank exploded when they were most of the way to the moon, when the life-support needs were greatly reduced compared to those at the beginning of the mission. And there is no reason that it couldn't have happened on the way back instead, in which case they wouldn't have had the LEM as a backup. Had that been the timing of the event, that crew would have died, regardless of how much of an option Gene Kranz said failure wasn't. It's worth noting that this also highlights just how big a gamble NASA took with Apollo 8, which had no LEM to use as a "lifeboat."

While many mistakenly blame President Richard Nixon for ending the Apollo program, the decision to end it actually occurred even before the space race was won, in 1967, during the Johnson administration. That was when production of the hardware for a long-term program started to end, resulting ultimately in only six actual successful lunar missions (Apollo 13 would have been a seventh). While part of the reason for ending it was the high costs to a nation laboring from the increasing expenses of Vietnam and Johnson's Great Society social programs, there was also a fear that eventually we would kill a crew if we continued for long, with no clear goal other than to beat the Russians in a race that they pretended not to be running. Apollo 13, while it was a rare moment for the nation to come together with the rescue, only reinforced this earlier concern.

But before the program ended completely and after the last lunar landing, NASA did undertake one more hazardous series of missions with Apollo hardware, though not to the moon. In 1973, America's first space station, *Skylab*, suffered a failure on its launch, when the meteoroid shield was torn off from the aerodynamic forces, taking with it one of the two main solar panels while partially deploying the other prematurely. NASA had to nurse the crippled facility in a "hold" position that resulted in an increase of heat within due to the loss of the shield. The temperature reached 125° F, and it was unknown whether or not the cabin atmosphere was breathable, due to potential toxins from outgassing of overheated materials. But the agency hastily planned a risky repair mission to be performed by the station's first crew—Pete Conrad, Paul Weitz and Joe Kerwin.

After rendezvousing with the station, the repair started with an open-hatch spacewalk from the Apollo capsule, with Weitz leaning out with a pole, his legs held by Kerwin. However, his attempt to release the stuck solar panel was unsuccessful. The crew then docked to the facility after several failed attempts, and entered after verifying that the air was breathable, albeit hot. From inside the airlock, they deployed a parasol to shield the structure from the sun, and the temperature finally started to come down, though the power remained low due to the missing and stuck solar arrays.

The improvised sun shade and single remaining main solar panel of Skylab

A couple of weeks into the one-month mission, Conrad and Kerwin once again put on suits and went out the airlock to attempt to free up the array. This time they succeeded, but it was almost a disaster, as Conrad was flung by the sudden release of the system after he removed debris from the hinge that was holding it in place. Had he not been tethered, he would have died as the life-support systems of his space suit were depleted, with him unable to get back to the spacecraft. But he was, and the two made it back in. The station was now almost fully functional, and would go on to host two more crews, giving the U.S. valuable long-duration space experience unmatched until the ISS was permanently crewed almost thirty years later. But with the end of *Skylab* in 1974, and the Apollo-Soyuz Test Project (ASTP) in 1975, the Apollo program was finally over.

The question was, what would come next?

Chapter 4

The Age Of Reusability?

What came next was the Space Shuttle and, in its development, it ostensibly drew on the lessons learned from the sixties programs. Clearly, as the famous science-fiction writer Arthur C. Clarke once wrote, we could not afford to continue to carpet the seabed of the Atlantic with launch hardware if we were to make space affordable, so the Shuttle should be reusable (though due to budget constraints, it ended up  only being partially so). But it was also meant to have a high flight rate, and make space routine, which implied a higher level of safety than had been attainable in Mercury, Gemini or Apollo.

And in terms of safety, the Shuttle ended up being a weird mishmash of the "human rating" requirements derived from the earlier years when we were putting men up on munitions. For instance, it carried along a lot of the same extensive (and expensive) documentation of components and processes (with various procedures often requiring multiple signatures, which paradoxically often meant that there was less scrutiny by individual signers who too often assumed that someone else would catch any problems), but it abandoned one of the primary requirements—zero-zero abort—because it would have incurred too large a cost and payload penalty. Thus, by NASA's own definition, the vehicle was *never human rated*.

While there were a number of abort scenarios during ascent, none of them became available until after the Solid Rocket Boosters (SRB) burned out about two minutes into the flight, because they could not be jettisoned while burning, and there was no thrust termination system in them. So once the solids were lit, the vehicle was going to go somewhere, and the crew was going to ride it until they burned out. And even after SRB shutdown and separation, abort was still another case of "attempted suicide to avoid certain death." For instance, I never met anyone in my experience in the industry who thought that the Return-To-Launch-Site (RTLS, the first available abort sequence) had a very high chance of succeeding. Trans-Atlantic Abort (TAL) to places

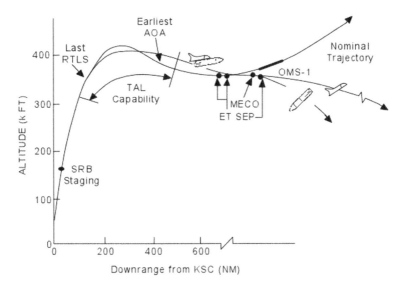

Space Shuttle Launch-Abort Scenarios
SRB: Solid Rocket Booster RTLS: Return To Launch Site TAL: Trans-Atlantic Abort
AOA: Abort to Orbit ET SEP: External Tank Separation MECO: Main Engine Cut Off

such as Rota, Spain, and Abort Once Around (AOA) seemed more viable and there was once an actual Abort To Orbit (ATO) when a main engine shut down prematurely due to a faulty sensor (which resulted in a lower orbit than planned due to the increased gravity losses from the reduced thrust, causing the failure to perform at least one planned experiment).

But of course, all of these abort scenarios depended on an intact orbiter. If the orbiter itself was broken, the crew was SOL[1],

---

1 Sort of Out of Luck. Some use a different word for the "S" in the acronym.

other than on the first four "test"[2] flights, in which there were two ejection seats for the two crew members, though they were useful only after the orbiter had separated from the rest of the stack. In fact, on the very first flight in 1981, commander John Young said afterward that if he had been aware of the amount of damage to the body flap and hydraulics from the

STS-1 Crew
John Young and Bob Crippen

initial overpressure of launch during the flight, he would have ordered an ejection of the crew once he got to a safe altitude after SRB burnout.[3] This would have resulted in the loss of *Columbia* then (rather than in 2003) and a much different history for the program, though it probably wouldn't have immediately ended it.

This cost saving was partially rationalized by analogizing the vehicle to an airliner. After all, no one on board has a parachute, because the cost and weight they add aren't worth the reduction of risk, given that few would even know how to use one, or have time to get out in an emergency. And it would make no sense to issue one to the crew, because they're the best hope the passengers have of getting the plane down safely (not to mention the psychological effect of seeing them get on board with the parachutes—everyone would be thinking, "What do they know that we don't?").

But to accept that analogy, one must presume that the reliability of the vehicle is very high, and so NASA did, right up to the morning of January 28th, 1986. That morning was atypically cold for Florida's Cape Canaveral. Icicles were hanging from the launch gantry and the air temperature was below specifications

---

2 The Shuttle was declared "operational" after the fourth flight on July 4th, 1982, but all this really meant is that the ejection seats were removed, and larger crews got to take their chances.

3 Oberg, James, as quoted by Alan Boyle, *MSNBC*, http://www.msnbc.msn.com/id/12243173/ns/technology_and_science-science/t/cosmic-log-april--/#. UEkGCB9WP4c

for the SRBs. But under schedule pressure, the decision was made to launch *Challenger* regardless.

Unknown at the time, and making conditions worse, was that prior to launch, a wind had been blowing across the cryogenic liquid-hydrogen tank, further chilling the already cold air that then blew across the right SRB and dropped its temperature even lower than ambient. The result was that the normally resilient rubber O-rings that seal the joints between the segments of the SRB became stiff.

At ignition of the main engines, the entire Shuttle assembly—which is literally bolted to the launch pad at the base of the SRBs—experiences a "twang" during which the entire stack bends in response to the off-axis thrust. The resulting stress on the SRBs affects the segment joints causing them to momentarily readjust. Then when the SRBs ignite, the stress is compounded as the SRBs momentarily balloon from the sudden pressure spike of combustion, further expanding and separating the joints. Normally, the O-rings comply with all this movement and maintain a tight seal. But this time, because of the lack of flexibility, a lower seal in the right booster didn't reseat properly after ignition and the combustion gases flowed past and destroyed the unseated seals. The joint was temporarily resealed by aluminum oxides from the burning solid propellant as the vehicle left the pad. But on that day, the side forces from wind shear aloft was higher than normal and the stresses reopened the joint as *Challenger* ascended through the atmosphere. The relentless hot gas quickly forced itself through the joint on the side facing the External Tank (ET) and acted like a blow torch[4] aimed directly at the lower support strut that connected the right SRB to the ET.

A little over a minute into the flight—about half a minute after maximum dynamic pressure (max Q), and less than one minute before the SRBs were scheduled to burn out and separate—the hot jet had weakened the lower strut to the point of failure and it

---

4 The jet of escaping hot gas generated a small amount of sideways thrust that caused the main engines to gimbal slightly to compensate and keep the overall thrust vector through the vehicle's center of mass. At the time, though, no one noticed this.

Crew of *Challenger* STS-51-L (from left to right): Ellison S. Onizuka, Michael J. Smith, Sharon Christa McAuliffe, Dick Scobee, Greg Jarvis, Ron McNair and Judy Resnik

broke. With that, the bottom of the right SRB forced itself away from the ET as the top of the SRB, now connected only by the upper strut, pivoted inward and ruptured the liquid oxygen tank at the top of the ET. At the same time, the hydrogen tank below was pushed forward by the thrust of the main engines into the oxygen tank above, rupturing it as well. The released propellants mixed in a huge fireball as the entire launch stack was torn apart. The two detached SRBs continued on independent upward trajectories until they were destroyed a few seconds later by a command from the range, while the orbiter, which lacked the ability to control its attitude, broke into pieces by aerodynamic forces that it was not designed to sustain. The intact crew module, with surviving crew, separated from the rest of the vehicle and fell from a dozen miles up into the Atlantic. All aboard were killed.

Prior to that morning, it had been an article of faith among upper management both at NASA and the contractors that Shuttle reliability was several nines (that is, 0.9999… chance of not having a catastrophic failure).

But those of us in the trenches (I was working for the prime contractor, Rockwell International in Downey, California at the time) knew that was nonsense. We'd all seen the fault trees that described all the things that could go wrong, and we could all assign our own probabilities to the various events, and know that we were sure to lose a vehicle (and crew) if we ever approached the planned flight rate of twenty or more per year (that was another dream for the program that ended when *Challenger* broke up off the coast of Florida). But any attempt to tell someone in upper management this was met with the same reaction as telling a proud parent that their kid was ugly, because most of those people were the same people who had developed the system in the seventies (most of them Apollo veterans). So we weren't shocked that we lost an orbiter that day, though we were surprised by the failure mode—that one hadn't been on any of the fault trees. We would have expected a main-engine (SSME) explosion or (as later happened to *Columbia*) a failure of the hardened carbon leading edges of the wings on entry, or damage to critical protective tiles.

In fact, Richard Feynman, the legendary physicist who was a key member of the Rogers Commission that investigated the *Challenger* disaster, noted this disparity himself:

> It appears that there are enormous differences of opinion as to the probability of a failure with loss of vehicle and of human life. The estimates range from roughly 1 in 100 to 1 in 100,000. The higher [probability of loss] figures come from the working engineers, and the very low figures from management. What are the causes and consequences of this lack of agreement? Since 1 part in 100,000 would imply that one could put a Shuttle up each day for 300 years expecting to lose only one, we could properly ask "What is the cause of management's fantastic faith in the machinery?"[5]

Part of the cause is that it was organizationally necessary to believe it, or declare the program a failure, not because it wasn't safe *per se*, but because it implied that the program was unaffordable.

That is, it is important to understand that the philosophy behind the Shuttle *did make sense*, despite false lessons learned

5 Feynman, Richard, "Feynman's Appendix To The Rogers Commission Report On The *Challenger* Accident," http://science.ksc.nasa.gov/shuttle/missions/51-l/docs/rogers-commission/Appendix-F.txt

in the wake of the *Challenger* and *Columbia* disasters (more on that to come). The flaw was not in the philosophy, but in the *implementation*, and much of it arises from our irrational approach to spaceflight safety.

Regardless of whether it carries crew or cargo, it is economically insane to build a reusable vehicle that is unreliable.

Consider an airline manufacturer who made airplanes that were perfectly safe. That is, in the event of a problem in flight, there was some mechanism (e.g., an ejection seat/canister/ whatever) for each passenger that returned them safely to the ground with very high reliability. So safety isn't the issue. But in order to make the development of the system affordable, it had to skimp on margins, redundancy and reliability, so that every third flight or so, an aircraft was lost. Even considering the cost savings from reliability reduction, this would still mean a loss of billions of dollars a year worth of aircraft for any reasonable traffic model (not to mention the loss of future business of passengers less than gruntled by a one-in-three chance of ending up in a corn field or bobbing in a lake instead of their planned destination). Is this a viable business model for a transportation company?

If this seems too bizarre a thought experiment, take a more reasonable one. Imagine that you are building aircraft for cargo only, with ejection seats for the crew. In fact, imagine that it doesn't even carry valuable cargo, just (say) commodities like grain that are cheaply replaceable. Would you say that, because there are no passengers or valuable cargo, the aircraft doesn't need to be reliable, that it's OK to lose a hundred-million dollar airplane occasionally?

The same was true for the Shuttle. While many mistakenly thought that it had quadruple redundant computers because it was a manned vehicle, the presence of crew was irrelevant to the need for its reliability—it had to be reliable because they cost about *two billion dollars each*, and they *only had four of them* at most, and sometimes only three operational at any given time, with production lines for components shut down in the eighties. The fact that it wasn't reliable to "n" nines (where "n" was at least three, or would fail less than one in a thousand times) meant that the program was in fact a failure.

Ultimately, this was why it was canceled, not because it killed fourteen astronauts who are, in comparison to just four Shuttle orbiters, (and at the ever-present risk of sounding cold hearted) certainly available in surplus numbers. After *Challenger*, the crew didn't need a fire-pole escape system, or ejectable crew capsule, or any of the other impractical changes that some wanted to make to improve the Shuttle's safety, any more than airline passengers need parachutes. What they needed was a *more reliable vehicle*. And in the end there was no way to provide it with the Shuttle, because its reliability was too compromised by its intrinsic design which resulted from its pinch-penny development. That's why the program failed and was canceled, not because it was unsafe *per se*. We simply couldn't afford to continue to lose orbiters.

There is a corollary to the rule that a reusable vehicle must be highly reliable to make economic sense. The corollary is that, by definition, such a vehicle would be sufficiently reliable to carry valuable cargo, including humans. This is simply because the vehicle itself is so costly to replace, regardless of the payload.

There is nothing about space vehicles that changes this intrinsic economic argument. Note that this is almost *completely independent* of what we think a human life is worth (unless we believe that the value is infinite, and many behave as though it is). This is why "human rating" (the modern politically correct version of "man rating") is an archaic relic of early and unreliable expendable vehicles, and is a phrase that should be banished from the policy and engineering lexicon.

Leaving reusable vehicles aside, even expendable launch vehicles are designed to be highly reliable today. This is not (just) because they might carry humans, but because they carry payloads that can cost billions of dollars and those payloads can be worth even more than that in terms of their potential revenue or as national assets. It's just bad for business to dump them in the ocean. In fact, the American launch vehicles that have been developed in the last couple of decades do seem to be quite reliable. United Launch Alliance's (ULA) Atlas V hasn't had a failure in

ULA Delta IV ULA Atlas V SpaceX Falcon 9

many flights, and neither has ULA's Delta IV (though it has much less of a record). The SpaceX Falcon 9 had only flown five times as of early 2013, but it had no primary mission failures , and it had at least reached a point at which it didn't seem to have any intrinsic design problems.[6]

But many in Congress, and within industry itself, don't seem to understand these fundamental economic issues, and continue to irrationally and arbitrarily think that something magical happens when passengers or crew are involved.

The history of space-disaster response provides numerous examples of false lessons learned in this regard.

It started with the *Challenger* disaster in 1986, in which the first response was to no longer allow commercial payloads to be carried by the Shuttle, unless there was no other way to perform

---

6 In early October of 2012, a year or so before publication of this work, the vehicle lost one of its nine first-stage engines on its first operational flight to the ISS for cargo delivery. Because it was designed for engine-out capability, the ISS mission was successful, at least in terms of the Dragon successfully berthing to the station with its payload, but a secondary payload was delivered to the wrong orbit as a result, and its life was far shorter than planned, entering the atmosphere after just a few days.

It should be noted that that version of the Falcon 9 (pictured above, launched) was retired in 2013. A new one (pictured above, on pad), with upgraded engines, a stretched first stage, and other modifications, was test flown from Vandenberg Air Force Base in late September of that year. It successfully delivered the satellites to their planned polar orbits, but it had an issue with the restart test of the second-stage engine.

that mission (e.g., an experiment in a rack in the mid-deck or in a SpaceHab module in the payload bay).[7] This was actually a good policy, because Shuttle had from its inception retarded the development of commercial launch vehicles by allowing a taxpayer-subsidized system to undercut the cost of any potential competitor. This (for obvious reasons) inhibited investment in same, and the decision to end that policy did enable the development of the commercial launch industry as it exists today. But as is often the case, particularly with government, it was an example of a good policy occurring for a really dumb reason. The policy wasn't changed to create a commercial launch industry—it was changed because people said that astronauts *shouldn't be risking their lives for the mundane task of launching satellites.*

Think about that. People risk their lives transporting goods all the time, millions per day. For instance, truck drivers die in accidents dozens, if not hundreds of times per year. No one proposes that we automate the trucks to prevent this (or at least no one has seriously to date). Why? Because it would make transportation for all but the most precious goods unaffordable. What it is about astronauts that makes them more special than truckers? Their education and training level? Certainly that does give them a higher pay grade and value, but that much?

Of course, the counterargument to this is that unlike trucks that need drivers, it is unnecessary to have a crew on a satellite launcher. But while that's true today, with the vehicles of the day, it won't necessarily be in the future, and if someone comes up with a satellite-launcher design that employs (and even risks) crew to reduce the cost of operations, it would make no sense for them to abjure it simply because someone once said don't risk crew to launch satellites.

The irrationality doubled down after the *Columbia* loss, with dramatic and expensive effects on policy. Following on from the flawed philosophy of the *Challenger* lesson, the Columbia Accident

7 Note that as is often the case with knee-jerk government decisions, this was a response that wouldn't have prevented the event that prompted it. That flight was not carrying a commercial payload—it was delivering a NASA Tracking and Data Relay Satellite (TDRS).

Investigation Board (CAIB) came up with a new flawed explicit recommendation from its chairman, Admiral Gehman himself[8]: "Don't mix crew and cargo."

Every ship that sails, every aircraft that flies, every truck on the road, every train on the rails, mixes crew and cargo. Again, what is so unique about space transportation that the two should suddenly become intrinsically immiscible?

This is not to say, of course, that one couldn't come up with a sensible vehicle concept that was primarily a passenger system, or exclusively for cargo, but in neither case should it be based *a priori* on a false lesson from a single flawed example (the Shuttle). The problem with the Shuttle wasn't that it mixed crew and cargo, but that it had far too many requirements in general, some of them in conflict with each other, and to repeat, the real problem with it was that one of its fundamental requirements was reliability, and it failed to meet it (partly, admittedly, because of the culture that operated it; more on that to come).

So what were the dramatic and expensive resulting effects on policy?

8 Santucci, Dave, "NASA urged to separate cargo from astronauts," *CNN* September 4th, 2003, http://www.cnn.com/2003/TECH/space/09/04/sprj.colu.house.hearing/index.html"

Chapter 5

The Safety Irrationality
Of Constellation

The Ares I launcher concept, part of the Constellation[1] program meant to get NASA back to the moon, was a poster child for this flawed lesson learned from Shuttle. Prior to being renamed "Ares I," it was called the Crew Launch Vehicle. It was NASA's response to the new requirement to segregate crew from cargo (and a successful one had it ever flown, since its payload performance was so poor it could barely carry crew, let alone cargo). While you wouldn't know it when actually considering

Artist's concept of Ares I
Crew Launch Vehicle (CLV)

the design (more on that shortly), the ultimate requirement for the system was to provide the highest possible level of safety, and dramatic reduction of the probability of Loss of Crew (LOC). And apparently cost was no object in order to achieve that goal. When the program was initiated in 2005, it had an initial cost estimate to Initial Operational Capability (IOC), including the Orion capsule that was the actual vehicle needed to carry the crew, of about *thirty billion dollars*. By April of 2009 (a few months before it was canceled), cost estimates to IOC had ballooned to *forty-four billion dollars*, with NASA continuing to lobby for more funding for it.[2]

---

1 See Appendix A for a description of the Constellation program.

2 Block, Robert and Matthews, Mark, "NASA estimates for new rocket soar again," *Orlando Sentinel*, April 2nd, 2009, http://articles.orlandosentinel.com/2009-04-02/ news/constellation02_1_constellation-program-nasa-rocket

NASA claimed that it needed the rocket, because the Atlas and Delta, to which the Pentagon entrusted billion-dollar defense satellites, and which had a flawless record in recent years, weren't "safe" enough for their astronauts.[3]

In fact, even Mike Griffin, the NASA administrator who came up with Ares I in 2005, disagreed with that, prior to becoming the head of the agency. A report from 2008 describes his *volte-face*:

In the CBS News interview published on Friday, November 14th, [2008] Dr. Griffin defended his plan to invest roughly $10 billion in taxpayer funding to create the new Ares 1 [sic] rocket, rather than using the existing, proven , saying "our selection was based first and foremost on crew safety ..." But on May 8, 2003, Dr. Griffin testified to the U.S. House of Representatives Committee on Science, rejecting suggestions that EELVs were not safe enough for human spaceflight. Dr. Griffin declared that no additional precautions, beyond a safety abort system, were necessary.

GRIFFIN: What, precisely, are the precautions that we would take to safeguard a human crew that we would deliberately omit when launching, say, a billion-dollar Mars Exploration Rover (MER) mission? The answer is, of course, "none". While we appropriately value human life very highly, the investment we make in most unmanned missions is quite sufficient to capture our full attention. Logically, therefore, launch system reliability is treated by all parties as a priority of the highest order, irrespective of the nature of the payload, manned or unmanned. While there is no EELV flight experience as yet, these modern versions of the Atlas and Delta should be as inherently reliable as their predecessors. Their specified design reliability is 98%, a value typical of that demonstrated by the best expendable

---

3 In fairness, NASA also claimed that some of the development costs of the Ares I were a "down payment" on the Ares V heavy lifter that would use some of the same hardware, so it wasn't fair to allocate the entire cost to that program, but such a reallocation would be somewhat arbitrary (some estimates were that it only saved a couple billion), and even the costs remaining would still be amazingly high.

vehicles. If this is achieved, and I believe that it will be, and given a separate escape system with an assumed reliability of even 90%, the fatal accident rate would be 1 in 500 launches, substantially better than for the Shuttle.[4]

Since then, the vehicles have in fact built up a good reliability record, as noted above. But just how safe did it have to be?

In 2009, a committee headed by industry veteran Norman Augustine was convened by the Obama administration to address the budget and schedule issues with the Constellation. Figure 1 is a chart from a presentation as presented to the Augustine Committee by Joe Fragola of Valador Information Architects, who did the safety analysis for the Exploration Systems Architecture Study (ESAS) which resulted in Constellation. Note the horizontal

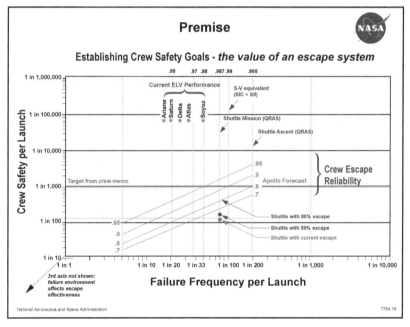

Figure 1. Safety goals for constellation were high in light of Shuttle experience
(Reproduced from NASA chart)

4 Space Frontier Foundation, "Will the real Mike Griffin please stand up?" *Spaceref*, November 19, 2008, http://www.spaceref.com/news/viewpr.html?pid=26969

red line—the target of a memo from the astronaut office was reportedly a one-in-a-thousand chance of injury or death during ascent. It was assumed that by combining a high launch reliability with a crew-escape system, this was an achievable and reasonable goal. Note also that the chart assumes in fact that the probability of LOC could be as low as one in seven thousand with an escape system. I would note as an aside that, while it's certainly worthwhile to solicit their input, it's not clear why a desire of the astronauts, who don't have to foot the bill, should have been taken as a requirement. We'll come back to that later.

Of course, even if NASA had legitimate concerns about safety, the tens of billions proposed for Ares/Orion were a dramatic misallocation of resources in the provision of it. The Vision for Space Exploration in general, and Constellation in particular, had the goal of returning humans to the moon. Considering all the things that can go wrong on such a mission—from ascent to lunar insertion, risking a solar storm en route, to landing on the moon, to taking off again from the lunar surface, to entry and recovery— given the decades of experience we have with launching things, ascent is one of the safest phases of the flight. Figure 2 shows NASA's own allocation of risk of loss of crew by mission phase. The pie represents NASA's estimate of likelihood of death or injury of an astronaut. For instance, they thought that the greatest

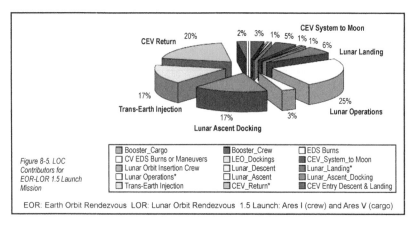

Figure 2. Ascent presents only 3% of total risk of loss of crew (LOC) for lunar mission

danger period was during operations on the moon itself—that is, given that a casualty would occur, they estimated a one-in-four probability of it occurring on the moon. Entrepreneur and engineer Jon Goff explained it at his blog in the fall of 2009:

[W]hen you're talking about exploration missions, ascent reliability is only a small component of the overall risk.

Since I've been doing a lot of rocket plumbing, let me use an analogy. Imagine you have a small rocket engine, like our igniter. It's got a tiny fuel orifice that's only a tiny fraction of an inch in diameter. Now, if the solenoid valve upstream of that orifice is small enough, it can meaningfully decrease the overall flow. But after you reach a certain valve size, the valve size ceases to be relevant to the overall flow rate. You could put a 1″ full port ball valve leading up to that orifice, and the flow difference between that and a solenoid valve with a 1/32″ port is going to be round-off error. Once the size difference between the most constricting component and the rest of the components gets past a certain point, you can pretty much ignore them.

There are tons of other analogies from electronics, manufacturing, structures, etc. Basically, in almost any system, you can improve one component only so far before you hit rapidly diminishing returns. A wise engineer will spend his resources in a way to maximize the overall system reliability, not just one small sub-component.

Unfortunately, that's exactly the mistake that the CxP guys have been making with Ares I. Even if, in spite of all evidence so far, and in spite of all historical precedent, Ares I really is as reliable as their Probabilistic Risk Assessments suggest, it still isn't a wise investment of capital when you look at the overall exploration mission.

Let's just go back to the math again.

(Quote continues)

> According to ESAS, the predicted probability of losing a crew
> on a lunar mission was something like 1.6% (or about 1/60).
> Of that, only about 1/2000 (or .05%) came from ascent risk–
> or about 3% of the overall crew risk for the entire mission. A
> mission that used the worst numbers they came up with for
> existing EELVs with an LAS [Launch-Abort System] attached
> estimated a 1/600 probability of losing a crew (or about .16%
> chance). That would increase the probability of losing a crew to
> 1.7% or about 1/59 …
>
> Investing tens of billions of dollars to reduce the probability
> of losing a crew from 1.7% to 1.6% only makes sense if there
> are no better safety investments out there. With a 1/600 ascent
> safety rating, about 90% of the danger to the crew is coming
> from other phases of the mission–which strongly suggests that
> the best safety return on investment is not in over optimizing the
> ascent reliability.
>
> Now, to be fair, Shuttle has a safety rating estimated to be around
> 1/100. At that reliability rating for crew launch, it would be one
> of the dominant crew safety risks in a lunar mission. Spending
> money to get it up past the 1/500 range is a good investment. But
> after a point, if your goal is to improve the overall probability
> of getting a crew safely to the moon and back, you're best off
> finding other areas to invest your money.[5]

As he notes, Mike Griffin's NASA was totally focused, budgetarily
speaking, on the first few minutes of the mission, whereas 90+
percent of the danger occurred afterward. No significant funding
was being spent on hardware needed to actually land on the moon
(e.g., a trans-lunar insertion stage, or lunar lander) because it was
all being eaten up by a "safer" rocket that the agency didn't need.

---

5 Goff, Jon, "Ares-I Ascent Reliability: Still Missing The Point," *Selenian Boondocks*
blog, October 9th, 2009, http://selenianboondocks.com/2009/10/ares-i-ascent-
reliability-still-missing-the-point/

And what's worse, even if the level of ascent safety desired was worth the money being spent, it was questionable whether or not the Ares I design was going to deliver it.

The idea seemed simple enough: take a Shuttle Solid Rocket Booster (SRB), and just put an upper stage on top of it. In fact, the motto of the program (and associated web site) was "Safe, Simple, Soon." The theory was that the SRBs had a good track record.[6] Part of the argument for the concept was that, because it used Shuttle (and later, Saturn-era) components, it was intrinsically "human rated," which only displayed an ignorance of what the phrase meant, since the Shuttle itself, as already described, was never human rated, and human rating, to the degree that it happens at all, is something that happens to an integrated system, not to individual components.

Independent analysis by the Aerospace Corporation indicated that NASA's methodology in risk assessment may have been flawed, and didn't take into account (for example) an explosive conflagration of the first stage, in which flaming gouts of burning solid propellant might ignite the parachutes of the aborting capsule. It also didn't consider the danger of a "hiccup" or final unpredicted burst of thrust from an SRB as its combustion tailed down at the end of the burn, potentially causing a rear-end collision with the upper stage during stage separation.[7] Nor did NASA take into account in its ESAS safety analysis the danger of resonant coupling with the structure and the SRBs during first-stage burn that threatened to shake the capsule, causing at a minimum difficulty to read instruments and a maximum of actual structural failure or crew injury. To ameliorate this was going to require the addition of a lot of cost and weight to the launch vehicle in the form of active dampers or other mitigations.

---

6 Ignoring the one that destroyed *Challenger* which, to be fair, had a failure that wouldn't happen in this particular vehicle configuration because, unlike the Shuttle, there would be no support strut to cut, or "twang" of the structure to stress the booster joints at ignition of the off-axis Space Shuttle Main Engines.

7 This isn't a problem on the Shuttle (or a Shuttle-derived vehicle) because the SRBs are ejected sideways after they burn out, with no risk of recontact even with thrust asymmetries.

There was another issue with safety. NASA decided to include a launch-abort system (LAS) with a launch tower on top of the Orion capsule. This was a very complicated system, involving upward-firing solid-rocket motors whose thrust was "turned" downward by bent nozzles at the top of the tower, and deployable canards (forward wings) to help control the trajectory after separation from the booster in an abort. It was a significant contributor to the cost, and including the blast protector shield on top of the capsule, it added about fourteen thousand pounds to the weight of the payload (though the actual payload penalty was less, because it wasn't intended to go all the way to orbit). I actually did some risk analysis as a subcontractor for the program, and we identified dozens of hazards that it introduced to the system. Moreover, many of these could give you a bad day *on an otherwise nominal mission.* As just one example, if you fail to jettison the system, which has to happen on every flight, you won't get to orbit, because there's too much weight for the upper stage, and you can't control the vehicle, or deploy chutes on entry with the thing stuck to the top. In other words, the mission fails, and the crew dies—because the safety device failed.

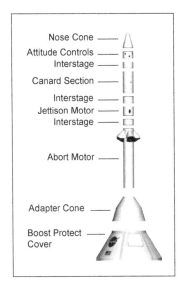

Orion Launch-Abort System

One would like to think that someone did an analysis adding up the hazards introduced by the system against the hazards that it ameliorates, and decided that it was worth the money and performance hit in terms of safety. I'd like to think that, but I have no reason to, and if it's been done, I've never seen the analysis. I think that it's more likely that it was a qualitative, not quantitative decision, based on the bureaucratic sense that no one wanted to have to go before Congress and explain why we lost a crew because they didn't have an abort system. The problem is, few think about the possibility of having to go before Congress and explain why the abort system killed the astronauts when there was nothing else wrong with the vehicle.

Their estimates of vehicle reliability were in fact based purely on Probabilistic Risk Assessments (PRA), which consisted of building fault trees of all the things that can go wrong and then assigning a probability of that particular event to come up with an overall reliability. They certainly weren't going to rely on actual flight history, because NASA's plan was to put crew on the vehicle after a single test flight (when its Bayesian—that is, purely empirical rather than analytical—probability of success, assuming the first flight was successful, was only two thirds, or 67%). Note that this is the same philosophy that they used for the first Shuttle flight, except in that case, it carried crew on the very first flight, because it wasn't designed to fly without them.

Why only a single test flight? Because the vehicle was so expensive to build and fly, that they couldn't afford a series of flight tests (to date this has been the bane of spaceflight cost and reliability, due to the high marginal cost of expendable rockets).

And how much confidence can we have in NASA's analyses? Well, Fragola's estimates are not inspirational. Some members of the Augustine Committee, particularly Jeff Greason, co-founder and CEO of XCOR Aerospace, and chairman, Norm Augustine, clearly were skeptical in a hearing in Huntsville, Alabama in August of 2009:

Jeffrey Greason: There is something that I think that we have come to understand as we've heard a lot of fact-finding briefings but what I want to get on the record because people who are listening to these briefings may have not been through all of that process which is, you really have to be careful about comparing the probabilistic risk assessment of an as-yet unflown vehicle to the demonstrated reliability of flown vehicles. The probabilistic risk assessments look at hardware-driven random failures and then you have a very sophisticated methodology in which you do your very best to look at what that effect is on the system failure

(Quote continues)

> but in wild [sic] terms historically, about 10% of the failures of launched vehicles are driven by those kinds of random effects. So, not only do PRAs grossly overstate the reliability of an as-yet unflown system, a fact everybody is aware of, but it also means that you have to be really careful driving your program design with factors of 50% or factors of two on probabilistic risk assessments because at the end of the day, *you don't really know what factor with a real reliability launcher is within a factor of ten* [sic]. So there is nothing wrong with using PRA to guide your decisions but you've got to use the numbers with great caution. [Emphasis added]

Joe Fragola: I think that is a very important point. And it's one of the strengths of the Ares I because the demonstrated reliability of the SRB is solid in more ways than one. And all the other vehicles also require a second stage. So the second stage to a degree is not a discriminator on safety because all the vehicles require a new second stage. The first stage of all the vehicle alternatives with the exception of the Shuttle C, the demonstrated reliability and the demonstrated risk is much, much better for the Ares I. From that perspective, the Ares I is far superior from all the other alternatives. If you look at the Delta IV Heavy for example, we have only had nine Delta IV launches and only I guess two Heavy launches and in the first launch, there was a discrepancy.

Okay? So from precisely the perspective you are speaking, that is one of the strengths of the Ares I.

Jeffrey Greason: But, let me add that your history will ... you're taking a big risk by assuming in that statement that future new launch vehicles, however, derived will not also experience early anomalies because that is the history of ...

Joe Fragola: Absolutely. If you look ... it will take me a long time go through the whole thing, but yes, absolutely. Were you to look into the history of a launch vehicle it is very important

(Quote continues)

how many test flights you have. The growth history of any launch vehicle is significant as any other developmental system and that has to be taken into account. When we looked at these numbers here, these are numbers that in my comparative chart, when number is taken at the mature level of all the launch vehicles, at the first lunar flight, we call it. So in other words, we anticipated successes on the Delta IV Heavy until 2015 and still the Ares I forecast is two times better. And it is precisely because of the failure modes interacting with the abort effectiveness. Yes, sir.

Norman Augustine: If I may, I would like to weigh on this too. Jeff said much of what I wanted to say but my own experience has been with calculated reliability has been not very happy [sic]. The one thing I have learned is that calculated numbers are always higher than the real world numbers, almost invariably. Over the years, I have kind of drawn the conclusion that the expendables are somewhere like 96% moving to 97%. Shuttle is like 98% moving to 99%. The Apollo, excuse me ... the Astronaut Office said I think, they wanted three 9s at 95% confidence and even if we use the calculated numbers that you've got, we don't get to that level by a considerable factor and it's likely the real numbers are going to be well under the calculated numbers ...[8]

First, in response to Fragola's comment, neither Atlas or Delta would have required a new second stage to be used for crew. But I emphasized Greason's point because a) it's correct and b) people like Fragola don't seem to believe it, or at least pretend not to. They are perfectly comfortable in quoting probabilities of loss of crew to four significant figures (e.g., 1 in 2183) when some of the factors used to develop the numbers are barely known to within an order of magnitude. One of the very first things that scientists and engineers are taught is that you can't get an answer more precise than the precision of the least precise factor from which

8 Transcript, public meeting of the Augustine Committee, August 7th 2009, Huntsville, AL, http://www.nasa.gov/pdf/378379main_080709_Huntsville_Transcript.pdf

it is derived. For example, we know the gravitational constant to many places, but when you multiply it by a local gravitational field that you only know to two places, that is the maximum precision that can be reasonably used to express the mass of that body. A good professor will mark down an answer on a student's test that, while accurate, is unjustifiably precise. When I see engineers doing the same thing, I tend to think that they're trying to impress the innumerate who don't understand the difference between precision and accuracy. And I think that the safety numbers for Ares I were precisely wrong.

One other important point is that PRAs cannot take into account the kind of management failures that resulted in the loss of both Shuttle orbiters. When one examines both cases, at heart, the reasons we lost them was actually not because of flawed designs, as unreliable as anyone might imagine them to be. We lost the Shuttles because of poor decisions by NASA management, and perhaps its very culture, driven by resource constraints that had been a hallmark of the program from its inception. In a speech to a group of Navy officers at the Naval Academy in 2005, Admiral Gehman, head of the CAIB, described the true cause of the failure:

> In the course of working closely with NASA engineers and NASA scientists as we tried to solve what had happened to the *Columbia*, we became aware of some organizational traits that caused our eyebrows to rise up on our heads. After not very long, we began to realize that some of these organizational traits were serious impediments to good engineering practices and to safe and reliable operations. They were doing things that took our breath away.
>
> We concluded and put in our report that the organizational traits, the organizational faults, management faults that we found in the space shuttle program were just as much to blame for the loss of the *Columbia* as was the famous piece of foam that fell off and broke a hole in the wing. Now, that's pretty strong language,

(Quote continues)

and in our report, we grounded the shuttle until they fixed these organizational faults.

I need to give you the issue from the NASA point of view so you can understand the pressures that they were under. In a developmental program, any developmental program, the program manager essentially has four areas to trade. The first one is money. Obviously, he can go get more money if he falls behind schedule. If he runs into technical difficulties or something goes wrong, he can go ask for more money. The second one is quantity. The third one is performance margin. If you are in trouble with your program, and it isn't working, you shave the performance. You shave the safety margin. You shave the margins. The fourth one is time. If you are out of money, and you're running into technical problems, or you need more time to solve a margin problem, you spread the program out, take more time. These are the four things that a program manager has. If you are a program manager for the shuttle, the option of quantity is eliminated. There are only four shuttles. You're not going to buy any more. What you got is what you got. If money is being held constant, which it is—they're on a fixed budget, and I'll get into that later—then if you run into some kind of problem with your program, you can only trade time and margin. If somebody is making you stick to a rigid time schedule, then you've only got one thing left, and that's margin. By margin, I mean either redundancy—making something 1.5 times stronger than it needs to be instead of 1.7 times stronger than it needs to be—or testing it twice instead of five times. That's what I mean by margin.

It has always been amazing to me how many members of Congress, officials in the Department of Defense, and program managers in our services forget this little rubric. Any one of them will enforce for one reason or another rigid standard against one or two of those parameters. They'll either give somebody a fixed budget, or they'll give somebody a fixed time, and they forget that when they do that, it's like pushing on a balloon. You push

(Quote continues)

in one place, and it pushes out the other place, and it's amazing how many smart people forget that.

The space shuttle *Columbia* was damaged at launch by a fault that had repeated itself in previous launches over and over and over again. Seeing this fault happen repeatedly with no harmful effects convinced NASA that something which was happening in violation of its design specifications must have been okay. Why was it okay? Because we got away with it. It didn't cause a catastrophic failure in the past. You may think that this is ridiculous. This is hardly good engineering. If something is violating the design specifications of your program and threatening your program, how could you possibly believe that sooner or later it isn't going to catch up with you? For you and me, we would translate this in our world into, "We do it this way, because this is the way we've always done it."

The facts don't make any difference to these people. Well, where were the voices of the engineers? Where were the voices that demanded facts and faced reality? What we found was that the organization had other priorities. Remember the four things that a program manager can trade? This program manager had other priorities, and he was trading all right, and let me tell you how it worked. In the case of the space shuttle, the driving factor was the International Space Station.

In January of 2001, a new administration takes office, and the new administration learns in the spring of 2001 that the International Space Station, after two years of effort, is three years behind schedule and 100 percent over budget. They set about to get this program back under control. An independent study suggested that NASA and the International Space Station program ought to be required to pass through some gates. Now, gates are definite times, definite places, and definite performance factors that you have to meet before you can go on. The White House and the Office of Management and Budget

(Quote continues)

(OMB) agreed to this procedure, and the first gate that NASA had to meet was called U.S. Core Complete. The name doesn't make any difference, but essentially it was an intermediate stage in the building of the International Space Station, where if we never did anything more, we could quit then. And the date set for Core Complete was February 2004. Okay, now this is the spring of 2001.

In the summer of 2001, NASA gets a new administrator. The new administrator is the Deputy Director of OMB, the same guy who just agreed to this gate theory. So now if you're a worker at NASA, and somebody is leveling these very strict schedule requirements on you that you are a little concerned about, and now the new administrator of NASA becomes essentially the author of this schedule, to you this schedule looks fairly inviolate. If you don't meet the gate, the program is shut down; they took it as a threat. If a program manager is faced with problems and shortfalls and challenges, if the schedule cannot be extended, he either needs money, or he needs to cut into margin. There were no other options, so guess what the people at NASA did? They started to cut into margins. No one directed them to do this. No one told them to do this. The organization did it, because the individuals in the organization thought they were defending the organization. They thought they were doing what the organization wanted them to do. There weren't secret meetings in which people found ways to make the shuttle unsafe, but the organization responded the way organizations respond. They get defensive. We actually found the PowerPoint viewgraphs that were briefed to NASA leadership when the program for good, solid engineering reasons began to slip, and I'll quote some of them. These were the measures that the managers proposed to take to get back on schedule. These are quotes. One, work over the Christmas holidays. Two, add a third shift at Kennedy Shuttle turnaround facility. Three, do safety checks in parallel rather than sequentially. Four, reduce structural inspection requirements. Five, defer requirements and apply the reserve,

(Quote continues)

and six, reduce testing scope. They're going to cut corners. That's what they're going to do. Nevertheless, for very good reasons, good engineering reasons, and to their credit, they stopped operations several times, because they found problems in the shuttle, and they got farther and farther behind schedule.

Well, two launches before the *Columbia*'s ill-fated flight—it was in October—a large piece of foam came off at launch and hit the solid rocket booster. The solid rocket boosters are recovered from the ocean and brought back and refurbished. They could look at the damage, and it was significant. So here we have a major piece of debris coming off, striking a part of the shuttle assembly. The rules and regulations say that, when that happens, it has to be classified as the highest level of anomaly, requiring serious engineering work to explain it away. It's only happened six or seven times out of 111 launches. But the people at NASA understand that if they classify this event as a serious violation of their flight rules, they're going to have to stop and fix it. So they classify it as essentially a mechanical problem, and they do not classify it as what they call an in-flight anomaly, which is their highest level of deficiency.

Okay, the next flight flies fine. No problem. Then we launch *Columbia*, and *Columbia* has a great big piece of foam come off. It hits the shuttle. This has happened two out of three times. Now, we go to these meetings. *Columbia* is in orbit, hasn't crashed, and we're going to these meetings about what to do about this. The meetings are tape-recorded, so we have been listening to the tape recordings of these meetings, and we listen to these employees as they talk themselves into classifying the fact that foam came off two out of three times as a minor material maintenance problem, not a threat to safety. Why did they talk themselves into this? Because they knew that, if they classified this as a serious safety violation, they would have to do all these engineering studies. It would slow down the launch schedule. They could not possibly complete the International Space Station

(Quote continues)

> on time, and they would fail to meet the gate. No one told them to do that. The organization came to that conclusion all by itself. They trivialized the work. They demanded studies, analyses, reviews, meetings, conferences, working groups, and more data. They keep everybody working hard, and they avoided the central issue: Were the crew and the shuttle in danger? [This was] a classic case where individuals, well-meaning individuals, were swept along by the institution's overpowering desire to protect itself. The system effectively blocked honest efforts to raise legitimate concerns. The individuals who raised concerns and did complain and tried to get some attention faced personal, emotional reactions from the people who were trying to defend the institution. The organization essentially went into a full defensive crouch, and the individuals who were concerned about safety were not able to overcome.[9]

Of course, this is all part of the entire reliability equation. But the two Shuttle orbiters did not fail because they were bitten by the "great unknown" or even the "known" (e.g., main engines blowing up). The first failed because the engineers were overruled by managers. And even then the failure was not intrinsically fatal. It was only the "bad luck" of the blow-by torching the SRB's support strut that destroyed the system. The second was due to the decision to continue to fly even though foam shedding was an ongoing problem. So, if they had not made a bad *management* decision about launching at colder temps than ever before and against the will of the engineers who understood the problem, and had the *managers* not continued to fly with a known safety problem in both cases, the Shuttle would have had a 100% success record.

---

9 Gehman, Harold, Admiral USN, Retired, "Ethical Challenges For Organizations: Lessons Learned From The *USS Cole* And *Columbia* Tragedies," Naval Academy Lecture, December 7, 2005, http://www.usna.edu/Ethics/publications/documents/GehmanPg1-28_Final.pdf

The Proposed Orion Capsule

So when it comes to reliability, the managers are a much more significant factor than is the hardware. And there is little reason to think that this would have been any different with Constellation, because its resources were just as constrained, if not more so than those for the Shuttle program. When Constellation was canceled by the Obama administration in early 2010, it was far over budget, and its schedule had been slipping more than a year per year, with little hope of flying before 2018. As a consequence of inadequate funding for both the Ares I launcher and the Orion spacecraft, redundancy, along with performance margins, was disappearing on an almost-monthly basis, and work on the planned heavy-lift Ares V had barely begun.

Although Constellation was a program that needed to be put out of space enthusiasts' and taxpayers' misery, in the fall of 2010 it was (unfortunately) partially resurrected by Congress. The Ares I launcher remains canceled. However, Orion was reborn as something initially called the "Multi-Purpose Crew Vehicle," and then later renamed back to "Orion." The heavy-lift Ares V lives on in the form of a new program called the Space Launch System (SLS) in which the program goal is to utilize Shuttle hardware to deliver 70 tons to low earth orbit (LEO) by 2017, and 130 tons to LEO by 2021.

The SLS, having been essentially designed in the Senate and having no clear mission requirements other than to maintain the existing Shuttle work force, has been pejoratively dubbed by many of its numerous detractors "the 'Senate' Launch System." And like Constellation, it is not being provided sufficient funds to achieve its goals on schedule, if at all. In fact, an independent cost assessment (ICA) on the project (as well as on Orion and another NASA program) was performed by the consulting firm Booz Allen

Hamilton, and in August of 2011, they released a report that included the following conclusions:

> Program estimates assume large, unsubstantiated, future cost efficiencies leading to the impression that they are optimistic. A scenario-based risk assessment, which excludes cost estimating uncertainty and unknown-unknown risks (historically major sources of cost and schedule growth), reveals ... reserves are insufficient.
>
> The Programs' estimates are serviceable and can be used for near-term budget planning in the current 3- to 5-year budget horizon. Beyond this horizon, the inclusion of large expected cost savings in the estimates, the beginning of development activities, and the potential for significant risk events *decreases the ICA Team's confidence in the estimates.*
>
> Individual Program cost and schedule estimates are managed separately and are not in alignment. Lack of cost and schedule alignment *makes it difficult to estimate the cost impact of schedule slips and vice-versa.* Non-integrated cost and schedule estimates are sufficient for trade studies, but do not facilitate ongoing baseline management.
>
> Program estimates *were shaped to fit within an anticipated budget profile.* While this is common practice, it is not consistent with GAO cost estimating best practices. Shifting costs to later years to fit upcoming budget caps decreases each Program's availability of funds for risk reduction, technology maturation, and *exposes the Programs to out-year cost growth.*[10] [Emphasis added]

---

10 Booz Allen Hamilton, "Independent Cost Assessment Of The Space Launch System, Multi-Purpose Crew Vehicle And 21st Century Ground Systems Programs, Executive Summary of Final Report," August 19th, 2011, http://www.nasa.gov/pdf/581582main_BAH_Executive_Summary.pdf

These are all related problems. The fourth point is perhaps the most revealing, and at the root of all of them. Normally, a development and operational program like this has a life-cycle cost profile: it starts small in the early years as initial design studies are performed; then ramps up as it goes into full development and later operations; then it tails back down in the out years as it is phased out and replaced by something else. As the old Monty Python skit goes, it's like a dinosaur, very thin at both ends and very large in the middle. And the program would be planned, and budget requests made, on that basis.

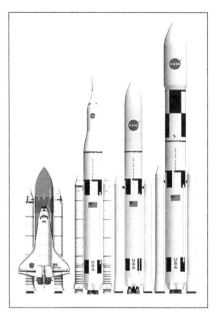

The Space Launch System (SLS) configurations (Space Shuttle for scale)

But that's not how Congress chose to fund it. Instead, the funding is basically relatively flat and steady from the beginning. So, NASA has been compelled to try to work within a budget profile that overfunds the project initially, and then starves it a few years later. The obvious implication is that the only apparent purpose for the program is to maintain the Shuttle work force; whether or not it succeeds at anything else is secondary. This also explains the lack of alignment of cost and schedule, or adequate reserves. Simply put, it doesn't matter, as long as the money is spent, and spent in the right states and districts of the authorizers and appropriators on the Hill.

It should be noted that NASA currently plans only two flights for the SLS—one in 2017 to demonstrate the 70-ton capability, and one with a crew in 2021, to . . . somewhere. They have said that, when operational, it may only fly every couple of years. What are the implications of that, in terms of both cost and safety?

Cost wise, it means that each flight will cost several billion dollars, at least for those first two flights. If, once in operation, it has a two- or three-billion-dollar annual budget (a reasonable guess based on Shuttle history), and it only flies every couple of years, that means that each subsequent flight will cost anywhere from four to six billion dollars.

From a safety standpoint, it means that its operating tempo will be far too slow, and its flights far too infrequent, to safely and reliably operate the system. The launch crews will be sitting around for months with little to do, and by the time the next launch occurs they'll have forgotten how to do it, if they haven't left from sheer boredom to seek another job.

As a last-ditch effort to try to preserve the Shuttle in 2010, some suggested that it be maintained until we had a replacement, but to fly it only once per year to save money.[11] The worst part of such a proposal would have been the degree to which the system would have been even less safe, given that it was designed for a launch rate of at least four flights per year. It was unsafe to fly it too often (as NASA learned in the 80s as it ramped up the flight rate before *Challenger*), and it would be equally so to fly it too rarely. NASA's nominal plans for SLS compound this folly, which is magnified by the fact that both internal NASA studies and independent industry ones have demonstrated that there is no need for such a vehicle to explore beyond earth orbit (existing launchers could do that job just fine, with orbital mating and operations), and it is eating up all the funding for systems, such as landers and orbital propellant storage facilities, that *are* necessary. All of this is just more indication that actually accomplishing things in space is the lowest priority for Congress (and unfortunately, the space agency itself, otherwise, the administrator would be more honest with the appropriators on the Hill).

In late 2012, the Space Foundation released a report that, among other things, criticized NASA's risk-averse and (perversely) dangerous safety culture, describing generally some of the

---

11 This demonstrated that they didn't understand the program costs, since the annual cost of operations of the Shuttle weren't strongly dependent on flight rate.

(Quote begins)

problems that I have discussed here specifically, and reinforcing Admiral Gehman's anecdote:

> Looking back at the era of Administrator Goldin's "Faster, Better, Cheaper" missions, truly trying to be the best in all three of these aspects means that risk is probably going to be the area most vulnerable to sacrifices. It is not surprising that the "Faster, Better, Cheaper" philosophy ultimately proved to be riskier than NASA or Congress was willing to countenance. In addition, the tendency to view the world in a technological framework means that NASA often wants to maximize performance. If project managers try to maximize performance and minimize risk, then cost and schedule tend to be less tightly controlled, as has been seen on numerous programs.
>
> Bureaucracies usually try to mitigate risk by adding procedures and regulations to existing practices. This effort results in increased paperwork, overhead, and transaction costs that may ultimately outweigh the benefits of the regulation in the first place. For example, in the processing of the Space Shuttle's Solid Rocket Boosters, line workers proposed a procedural change that would speed up processing (and arguably make the process more reliable), but when they tried to introduce the change, they were told that it would be too expensive to change the applicable manuals and written procedures. The number of manuals and written procedures, in turn, arose from a desire to minimize risk by making sure everything is well documented. NASA ended up with a time-consuming, (potentially) *less safe procedure*, as an indirect result of behaviors intended to ensure safety.
>
> Adding safety systems and redundancy increases the complexity of a system and introduces entirely new components that can break or malfunction, potentially increasing risk. In addition, flagging and marking everything can overwhelm operators, who then start treating potentially legitimate problems as acceptable

(Quote continues)

> because they have not yet failed. In the U.S. space program, it is very typical to try to increase safety (or at least reduce risk) by spending money. Many NASA activities end up being planned almost as rigorously as human spaceflight. This introduces rigidity, increased transaction costs, and inefficiencies in areas where a purely technical approach to risk management is not appropriate. [Emphasis added][12]

As already noted, the Orion Launch-Abort System is a perfect example of the problem of the introduction of complex systems for the purpose of safety. But I would also add that a "purely technical approach to risk management" is essential if we are to ever make human spaceflight affordable, which includes a rational allocation of risk acceptance and resources.

So NASA's approach to safety in Constellation was doubly irrational. First, in elevating it to the highest value, and second, after doing so taking an approach that provided very little safety for the money.

This attitude toward safety also makes it impossible to even do useful trade studies. When I was consulting to a subcontractor on one of the competing teams for the Crew Exploration Vehicle (CEV, later renamed "Orion") in the mid-aughts, I was responsible for supporting these on various aspects of the design. A trade study is an analysis to determine the best concept to meet a given set of system requirements. For example, we traded landing on land versus landing in water, or landing with parachutes versus landing with rockets or airbags. But in order to perform such a trade, one has to first define what "best" means, and to quantify it. Typically, we do this by coming up with various evaluation criteria (e.g., cost, schedule, performance, safety, reliability, technology risk), and then assigning weights to them, to allow the "scoring" of

---

12 "Pioneering: Sustaining U.S. Leadership In Space, page 21, The *Space Foundation*, released December 4th, 2012, http://www.spacefoundation.org/sites/default/files/downloads/PIONEERING.pdf

one concept versus another. This is typically done by a Delphi or consensus method, such as the Analytical Hierarchy Process, which can be used to at least attempt to quantify subjective opinions.

There are two problems that arise in this process when we believe that "Safety is Number One." First, it implies that the greatest weight would be given to this criterion, even though clearly, cost and schedule, not to mention performance (ability to actually accomplish what you're trying to accomplish) are very important as well. But we did manage to overcome that, and come up with a set of weights, with the plurality going to the politically correct safety god.

The real problem arises when trying to actually score the various concepts. Quantifying cost and schedule is obviously straightforward, as is performance if it's (say) weight, or payload capability. Quantifying the degree to which something is "safe" is much harder. The difficulty is magnified, if not rendered impossible, by the culture of the Safety and Mission Assurance (S&MA) community in the space industry, which believes, in the face of all human history and reality, that "safe" and "unsafe" are absolute states, rather than degrees along a continuum.

So we'd go to the S&MA folks, and say, "Can you tell us which is safer, Concept A or Concept B? Can you assign a value of, say, 3 to one, and 7 to the other?" And the response was "No. That's not how safety works. It's either safe or it's unsafe, and if it's unsafe, we don't fly until we fix it and make it safe. Period." I had much more hair before I started that project than afterward, so much did I pull in frustration.

Unfortunately, at least with regard to the elevation of safety as the highest value, NASA is continuously pushed into it by irrational worship of the safety totem by Congress. One recent example occurred in the spring of 2011 when, in reference to the successful penultimate Shuttle flight, Congressman Chaka Fattah (D-Penn), ranking member of the House Appropriations subcommittee for the agency, said to the administrator that he was "grateful that safety was your number one priority for the crew."[13] All this does is point out how unimportant are other things, such as actually accomplishing things in space.

A few months later, this irrational obsession with safety was on full display in deliberations about what to do with the International Space Station if it couldn't be reached by the Soyuz.

---

13 "NASA Appropriator Cheers NASA Administrator on Successful Mission Launch" *PRNewswire*, May 17th, 2011, http://www.prnewswire.com/news-releases/nasa-appropriator-cheers-nasa-administrator-on-successful-mission-launch-122043924.html

Chapter 6

Our Irrational Approach
To ISS Safety

The International Space Station

For over a dozen years now, the International Space Station (ISS) has never seen a day in which it didn't have occupants. Many thought when it was first permanently crewed, back in November of 2000, that it was a watershed in history—the day after which there would never again not be humans living off the planet. That was certainly the plan, because it was assumed that there would be follow-on programs even before the ISS was decommissioned. But it almost became another false start. As a result of the failure of the upper stage of a planned Russian Roscosmos Progress cargo flight to the station in August of 2011, NASA was actually contemplating, at least temporarily, giving up our tentative first foothold in the long climb to eventual space settlement.

Soyuz Launch

The problem was that the agency could no longer rely on the Russians' crew launcher, because it is essentially the same rocket as the one used for that failed Progress flight. Until the Russians had determined what caused the problem and how to fix it, their rockets could not be trusted. The implications for the ISS were potentially dire. First, there were plans to deliver a new crew on the Soyuz a few weeks later, which had to be delayed.

Also delayed was the return to earth of three crew members at the time, which had been planned about the same time as their replacement would be arriving. It wasn't an immediate issue, other than prolonging their stay in space, with whatever health detriments might accrue. The real problem was that they couldn't delay it indefinitely, because the Soyuz capsule only has a limited on-orbit life of seven months, after which it cannot be used to return crew home with confidence. The one they planned to bring back would have reached its use-by date in late October. So they would have had to come home then, leaving only three aboard, reducing or eliminating any science that could be performed there until it reached a full crew complement again.

Soyuz TMA-7 Spacecraft

The remaining three could in theory have stayed until January, when their Soyuz also would have started to get stale, but there were earthly issues to deal with in terms of their schedule. If they waited that long, they would have come down in a brutal Kazakh winter.[1] If they wanted to get in before that, the

---

1 In theory they could have come down in some other location, but the logistics of not landing in their normal landing site may have been considered untenable.

last time they could have left and still come down in daylight was in November. So for safety reasons, if the Russians couldn't figure out the problem (and get a new expedition up) NASA thought that they would have had to *abandon the station entirely* until they could. The agency believed that it could continue to maintain the facility remotely, at least for a while, but

SpaceX Dragon Capsule

it would have been a major psychological setback, particularly given that the station had just completed assembly and was about to start finally doing some serious research, including experiments with potential implications for medical cures. And it carried with it the very real possibility of something going wrong while uncrewed that could result in the loss of the station itself.

Ironically, it could have also been a setback for the most promising near-term means of reducing or eliminating our reliance on the Russians.

At that time, Space Exploration Technologies, Inc. (SpaceX) was scheduled to launch its Dragon capsule to the ISS in late November, with plans for a test docking with the system in early December (almost exactly a year after its first successful flight). But if there had been no one at the ISS in late November, there would have been no one to grab and berth the Dragon, or open the hatch to it after docking, so the mission would have had to be postponed until such a time as the station was occupied again.[2]

As it turned out, the Russians did figure out the problem, to NASA's satisfaction, and they abandoned plans to abandon the ISS. But it is illustrative of the kind of irrational decisions the agency will consider in the name of crew safety.

Consider, even if the Russians had been delayed longer, NASA didn't really have to abandon the ISS. In reality, they had several

---

2 This later turned out not to be an issue, because the flight was delayed for six months, to May of 2012, for other reasons.

options. The safest (but risky) solution would have been to fund a crash program to automate docking with the Soyuz (the Progress robotic spacecraft already has this capability), load up a fresh crew module with payload, but without crew, and launch it to replace the one that was going stale, thereby extending the stay of existing crew. Or, while it wouldn't be prudent, they could have hoped that what happened in August was just an anomaly, and gone ahead with the next Soyuz as planned. Or the astronauts could have taken the risk of a winter landing. Or they could have landed somewhere else on earth where the weather was better.

That NASA didn't seem to be considering any of these things, and was instead contemplating abandoning our only orbital outpost on which we've spent tens of billions over decades, even if only temporarily, speaks eloquently about our national perception of its importance, and trivializes it. It was essentially saying that—unlike commercial fishing, coal mining, construction, and liberating peoples—opening up frontiers, even the harshest final one, isn't worth the risk of human life, even a small number of them.

Aircraft carriers and other Navy ships cost billions, and their crew are expected to risk their lives to preserve the ship. I'll bet that there are plenty of people in the astronaut office who'd have been willing to take that risk, and in the unlikely event that there weren't, there are plenty of people fully qualified who are. It's what our ancestors would have done and how they created this great nation that once put men on the moon. But NASA was willing to both suspend research and to risk the loss of the station itself rather than hazard the lives of an astronaut crew.

But I would go further and argue that a "lifeboat," at least as defined by NASA—a vehicle that allows complete evacuation of the ISS all the way to earth—is an unjustifiable luxury. Consider the potential situations: something happens at the station that presents a hazard to the crew; or a crew member is injured or gets sick, and requires evacuation. Should the response be to bring *all of them all the way back to earth,* when it cost so much to get them into orbit in the first place? And does the "ambulance" scenario justify abandoning the station, with all the risks entailed to that facility? Are these reasonable requirements? Is a space facility, once again, for some irrational reason, special in this regard simply because it is in space?

Consider our research stations in Antarctica (the best existing analogue to the ISS). With its high winds and machinery-freezing temperatures (-100° F), combined with its altitude and thin air of almost ten thousand feet and continuous darkness, *Amundsen-Scott* station at the south pole is almost impossible to access during the austral winter. Between February and October, there have traditionally been no "ambulance" services for the few dozen people who winter over.

There have been several occasions when that they could have used one. There have been personnel emergencies, people have gotten ill, and one has died there.

In 1999 the station's winter physician, Dr. Jerri Nielsen, discovered that she had breast cancer. She actually did a biopsy on herself, and while it was impossible to evacuate her, there was a daring air drop of supplies needed for her chemotherapy (the equivalent for the ISS would be to send an emergency cargo mission). She survived until spring and was evacuated. She in fact survived almost a decade, but succumbed to a recurrence of the disease in 2009.

The following year, an astrophysicist fell ill from an unknown ailment and, despite advice and attempts at diagnosis via satellite, he died three days later.[3] Would it have made sense to evacuate the entire facility to ensure his rapid return?

In the austral summer of 2007 (at Christmas time), two men got into a drunken brawl, and were evacuated, because they could be. How would they have dealt with it had it happened in the winter?

In 2011, the communications technician came down with appendicitis. They performed an emergency appendectomy on him there, but if they could have medivaced him, they would have.

---

3 An autopsy after his body was returned to civilization revealed death from methanol poisoning, and some have speculated that it was homicide, but the ultimate cause remains a mystery.

A few weeks later, in late August, the station manager suffered a stroke, and there was no diagnostic equipment (e.g., an MRI scanner) available to determine the extent of the damage. Her family desperately wanted her evacuated, but she had to wait six weeks, until mid-October, before the early spring weather would permit a flight.

Despite all of these problems, one of them fatal (and Nielsen might have lived longer had she gotten better treatment sooner) there has never been a call by anyone to spend billions of dollars on a unique specialized emergency vehicle to provide 24/7/365 access to and from the Antarctic station, though given sufficient resources some clever engineers could probably come up with such a thing. And unlike NASA, the National Science Foundation has (sensibly) never gotten those kinds of resources. Because we recognize that sometimes it is worth taking risks for research, and that the lives of the researchers do not have infinite value, or even billions of dollars worth of value except, inexplicably, when it comes to space research.

In fact, almost as if to make the point, as these words were being written in late January of 2013, a Twin Otter logistics aircraft crash landed in Antarctica with three Canadians aboard. After a couple days, searchers gave up attempts to rescue them due to bad weather at the crash site, with winds exceeding a hundred miles an hour, and hazardous terrain where it went down. They are presumed dead either from crash injuries or exposure. Kelly Falkner, director of the National Science Foundation's Division of Polar Programs, said, "In many ways, their contributions make possible hard won but vital advances in scientific knowledge that serve all of mankind. Although everyone associated with the pursuit of science in Antarctica makes personal sacrifices to do so, very infrequently and sadly, some make the ultimate sacrifice."

Now of course, ignoring the ambulance issue, some will argue that there are greater risks to a space station than an Antarctic station—an equipment failure could create hazardous conditions, or it could be hit by debris or a natural object, breaching the hull to vacuum. And of course, it would be the first line of defense, a picket, in a space-alien invasion.[4] Whereas there aren't any natural hazards at the south pole as long as you keep your redundant equipment and fuel supplies stocked, and it's probably the last place on the planet that aliens would attack.[5,6]

It's true that the ISS is more hazardous than the *Amundsen-Scott* Antarctic Station with perhaps one of the greatest differences being that we have little to no experience dealing with medical emergencies, particularly surgery (for example, a burst appendix) in weightless conditions. On earth, even at the south pole, gravity can be relied upon to keep bodily fluids in place, but on orbit, a spurting artery could create a hazard to not just the injured or surgical patient, but the facility itself if floating blobs of blood were to get into the equipment. This really points out a need to develop techniques for such procedures, which NASA would prioritize if moving humanity into space were their priority.

But even given that, is the ISS more hazardous than (say) going into combat? Or working on an oil rig? Or in a coal mine?

Again, some will argue that there are times when ships have to be abandoned. That is why they have lifeboats.

And I agree with having a "lifeboat" for astronauts as long as it is a true lifeboat. That is, something that provides a way to get to another co-orbiting habitat, rather than NASA's ludicrous notion of evacuating all the way back to earth. NASA's approach is akin to having the *Titanic's* lifeboats built with the capability to get the passengers all the way back to Southampton, rather than to just provide temporary safety while awaiting rescue from the *Carpathia* or the *Californian*.

If NASA wanted to take the approach of evacuation to a co-orbiting habitat, I'd be all for it. But this obviously means having in-space infrastructure, such as additional co-orbiting facilities, and a transfer vehicle to get from one to the other. This way, if they had

---

4 At least until we get outposts farther out, perhaps at the Earth-Moon Lagrange points, or facilities in the outer solar system.

5 Unless they were trying to sneak up on us from below.

6 Yes, this is tongue-in-cheek (in case you were beginning to wonder).

to evacuate, there would be someplace they could go short of all the way back down the gravity well of earth, and be able to easily return once the situation was resolved. It would mean making—if not low earth orbit in general—at least the vicinity of the ISS, no longer a wilderness; a wilderness in which *Columbia* was doomed. I made this point in an essay a few days after *Columbia* was lost in early 2003, over a decade ago, that is very much in keeping with the topic of this larger work:

At the dawn of the 21st Century, Americans are coddled and take many things for granted.

We get into our cars, we drive out in the country or up into the mountains, and we expect to find gasoline, food at the grocery or general store, and a motel that will have bedding and indoor plumbing for our biological needs. If we're really adventurous, we'll not take a car, but instead a motor home, so that we can stock up on food and supplies and rough it out in the woods for a while.

But when our country was young, out on the frontier, there were no groceries. There were no conveniences. Sometimes, if one went too far over the verge, there weren't even the basic things that we needed to live, like water. Yet many went out into the wilderness, risking life and happiness, often for no reason than to see what was over the next mountain.

Let's move back into the 21st, or even the 20th Century, for a moment, and change the subject slightly (but only slightly, as we will see in a few paragraphs).

When a pilot takes off in an airplane, one of the fundamental things he does before spinning up the propeller and becoming airborne is to check out the aircraft. He walks around it, examining the control surfaces, the pressure in the tires, the fasteners that hold vital wings to critical fuselage. He tests the

(Quote continues)

controls and verifies that his manual activities result in aircraft response—rudder, aileron, elevator.

Then, he knows that the aircraft is ready for flight, and so is he.

Prior to each flight, the space shuttle undergoes the same procedure, except instead of a simple brief walk around by the pilot, it spends months under the tender ministrations of a division of troops, dedicated engineers and technicians, the "standing army" that claims so much of the cost of the system, to ensure that it is ready for its mission.

But consider: there are three phases to a space shuttle's mission.

The first is the launch phase, in which it is thrust out into the universe on a huge flaming tail of fire, briefly generating more power than the entire electrical output of the nation. We lost a shuttle during this phase 17 years ago, and everyone assumed that it was the most dangerous part of the flight.

The second phase is on orbit, in which the astronauts float, ethereally, accomplishing their mission. The sense of danger is almost nonexistent and palliated by the serenity of weightlessness and the silence of the emptiness of space, and beauty of the earth passing below, once every hour and a half.

The third phase is actually the most dangerous.

In this phase, the vehicle must reenter earth's atmosphere, and it must slow down by using the friction of that hypersonic air to drag it to almost the halt necessary for it to make final approach to the runway and land. It has an unimaginable amount of energy in orbit, and almost all of it must be dissipated into the thin gases at tens of miles of altitude, and (at least momentarily, until it can cool off) into insulating and heat-absorbing tiles on

(Quote continues)

the hottest portions of the structure, particularly the nose and leading edge of the wings.

The ascent environment, assuming that there are no catastrophic disassemblies of the stressed propulsion systems (as occurred on the final *Challenger* flight in 1986) is a cake walk compared to the entry, at least as far as the orbiter is concerned.

Yet prior to ascent, engineers spend months refurbishing and inspecting the vehicle, preparing it for launch. In contrast, prior to the much more strenuous descent, after having gone through the rigor of ascent, almost nothing is done, unless there's an obvious problem indicated by sensors. It is simply assumed that the ground preparation readied the vehicle for the entire mission, and that nothing will occur on orbit to make the return problematic.

Why? Because there's no capability in the system to do otherwise. There are no facilities in space to inspect or repair a shuttle orbiter. There are no tow trucks to rescue it if it has a propulsion failure. There are no motels to spend the night if they can't return on schedule. There are no general stores to purchase additional supplies of food—or air.

Every flight of a space shuttle (at least those that don't go to ISS) is a flight deep into the wilderness of space, in the equivalent of a motor home on which everything has to go right. Because, there's no other way home and delay is ultimately death—and "ultimately" isn't very far off.

I've written before about the fragility and brittleness of our space transportation infrastructure. I was referring to the systems that get us into space, and the ground systems that support them.

But we have an even bigger problem, that was highlighted by the loss of the *Columbia* on Saturday [February 1st, 2003]. Our orbital

(Quote continues)

infrastructure isn't just fragile—it's essentially nonexistent, with the exception of a single space station at a high inclination, which [because they were in different orbital planes[7]] was utterly unreachable by the *Columbia* on that mission.

Imagine the options that Mission Control and the crew would have had if they'd known they had a problem, and there was an emergency rescue hut (or even a Motel 6 for space tourists) in their orbit, with supplies to buy time until a rescue mission could be deployed. Or if we had a responsive launch system that could have gotten cargo up to them quickly.

As it was, even if they'd known that the ship couldn't safely enter, there was nothing they could do. And in fact, the knowledge that there were no solutions may have subtly influenced their assessment that there wasn't a problem.

The lesson we must take from the most recent shuttle disaster is that we can no longer rely on a single vehicle for our access to the new frontier, and that we must start to build the needed orbital infrastructure in low earth orbit, and farther out, to the moon, so that, in the words of the late Congressman George Brown, "greater metropolitan earth" is no longer a wilderness in which a technical failure means death or destruction.

NASA's problem hasn't been too much vision, even for near-earth activities, but much too little. But it's a job not just for NASA—to create that infrastructure, we will have to set new policies in place that harness private enterprise, just as we did with the

---

7 The difference in orbital planes at the time was about 90°, because of the relative positions of the longitude where the planes crossed the equator. Ignoring atmospheric drag and gravity losses, a 60° plane change requires as much velocity as it takes to get into orbit, and a 90° plane change takes about forty percent more than that (or about as much as it would take to achieve escape velocity from the earth's surface). So, even if there had hypothetically been a completely full External Tank attached to *Columbia* in orbit, and the main engines could have been restarted, there still wouldn't have been nearly enough propellant to match the orbital plane of the ISS.

(Quote continues)

> railroads in the 19th Century. That is the policy challenge that
> will come out of the latest setback—to begin to tame the harsh
> wilderness only two hundred miles above our heads.[8]

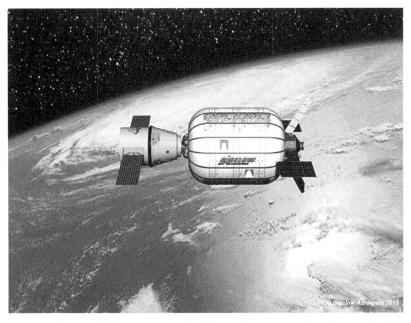

An alternative to NASA's return-to-earth "life boat" for emergencies that require ISS evacuation could be a co-orbital facility such as shown in this artist's concept of a Bigelow Aerospace expandable modular habitat (with a docked SpaceX Dragon spacecraft)

I think that providing such co-orbital facilities and transportation could be done for a small fraction of the cost that NASA has always been willing to spend on its "lifeboat." But it would have to be done commercially. For example, NASA could lease an expandable facility from Bigelow Aerospace, put it in the same orbit as the ISS, and modify a SpaceX Dragon to be a pure in-space vehicle which would have orbital refueling capability and, to minimize mass for its in-space excursions, it would have no

8 Simberg, Rand, "Into the Wilderness," Fox News, February 6th, 2003, http://www. foxnews.com/story/0,2933,77783,00.html

entry or recovery systems.[9] In addition to serving as an emergency safe haven for the ISS crew, the co-orbital facility could serve as a hotel from which short tours could be made to the ISS.

One way for NASA to make it happen as a commercial project would be for them to subsidize it, just as the Air Force provides funding to the airlines via the Civil Reserve Air Fleet to make their aircraft available in the event of an emergency need for transport (e.g., Federal Express moved much of the material for the first Gulf War). The fact that they never consider this and remain stuck in the current unaffordable paradigm of spending billions to devise a system to abandon a facility costing billions is just one more symptom of our irrational approach to safety when it comes to space. But I should note, to be fair, that the loss of *Columbia* did result in one brief fit of sanity on the part of NASA when it came to safety. We'll discuss that in the next chapter.

---

9 On January 16th, 2013, NASA and Bigelow announced a deal to mount an expandable Bigelow facility to the ISS to increase its habitable volume, though not for a separate co-orbital facility.

Chapter 7

The Hubble Repair Saga

Prior to *Columbia*'s fateful flight, NASA had planned to repair and upgrade the Hubble Space Telescope, which it could only do with the Shuttle. But the Hubble's orbit is in a similar orbit to that in which *Columbia* had been when it was destroyed on entry. That is, if a similar event had occurred during a Hubble repair mission, the crew and the ground would have been just as helpless, because there would have been no way to get to the ISS.

The Hubble Space Telescope in 2009

After *Columbia* didn't return, it was reported that NASA administrator Sean O'Keefe had been traumatized by having to be the person who told the families and loved ones on the tarmac at Kennedy Space Center that their husbands, wives, sons, daughters, friends weren't coming home. He obviously didn't want to have to do it again, and it appears that he became very risk averse. About a year after the disaster, and two days after President Bush had rolled out his new Vision for Space Exploration (which Constellation was ostensibly supposed to fulfill), O'Keefe announced the cancellation of the repair mission, largely on the grounds that it was too hazardous for the crew.[1] The only way to do it safely would have been to have another Shuttle ready on the pad while it was in orbit, for a potential rescue mission. It was also argued that they needed every available Shuttle flight to finish the ISS by 2010, per the new post-*Columbia* policy.

---

1 Britt, Robert Roy and Berger, Brian, "Hubble falls victim to changes at NASA," *NBC News*, January 16, 2004, http://www.msnbc.msn.com/id/3982359/ns/technology_and_science-space/t/hubble-falls-victim-changes-nasa/

The decision created a public-relations uproar. With the many spectacular pictures of distant galaxies, Hubble is one of the most popular things that NASA has done in the past couple decades, and the public was dismayed that it was going to be abandoned without a replacement (absent a repair mission, it probably would have failed within a very few years).[2] In early December, the National Academy of Science, in response to a congressional request, issued a report stating that the repair mission should be reinstated:

"The committee recommends that NASA pursue a shuttle [sic] servicing mission to HST that would accomplish the above stated goal. Strong consideration should be given to flying this mission as early as possible after return to flight," the report states.

Committee members acknowledged that while Hubble is important, they realized that NASA has other human spaceflight missions to deal with.

"We fully recognize that the International Space Station is a very high priority after return to flight," said committee member Richard Truly, a retired U.S. Navy Vice Admiral and director of the National Renewable Energy Laboratory in Golden, Colorado." After seven or eight flights, the technical configuration of the space station will be at a point that we concluded you could insert a Hubble flight."

The final report also rebutted O'Keefe's objections to Shuttle servicing as too risky, saying "the difference between the risk faced by the crew of a single Shuttle mission to the ISS—already accepted by NASA and the nation—and the risk faced by the crew of a Shuttle mission to HST, is very small. Given the intrinsic value of a serviced Hubble, and the high likelihood of success for a Shuttle servicing mission, the committee judges that such a

---

2 Foust, Jeff, "Life after Hubble" *The Space Review*, February 2nd, 2004, http://www.thespacereview.com/article/95/1

mission is worth the risk." Beyond that, the costs for the Hubble upgrade were already expended and the hardware, that would significantly improve the performance of Hubble, was already built and sitting in a warehouse waiting to be installed.[3]

The report again noted that the risk to the crew could be mitigated by having another "rescue Shuttle" waiting on a launch pad during the mission, and that the crippled orbiter itself could act as a safe haven for many days or a few weeks in a low-power mode until the rescue mission could occur. Of course, this wouldn't mitigate the risk of losing another third of the fleet when NASA still had to complete ISS assembly, nor did it address the possibility that the rescue Shuttle itself might fail. It also noted that a robotic servicing mission would be very high risk, in terms of mission success, given the state of the technology of the time. That a Shuttle flight was now perceived to be higher risk was only because of the experience of the *Columbia* loss.

O'Keefe resigned from NASA at the end of the year. Mike Griffin replaced him in the spring of 2005, and he had a different opinion about the Hubble mission. Hubble supporters were encouraged by his testimony in his confirmation hearing:

I believe that the choice comes down to reinstating a shuttle [sic] servicing mission or possibly a very simple robotic deorbiting mission. The decision not to execute the planned shuttle [sic] servicing mission was made in the immediate aftermath of the loss of Columbia. When we return to flight, it will be with essentially a new vehicle, which will have a new risk analysis associated with it and so on and so forth. At that time, I think we should reassess the earlier decision, and in light of what we learn after we return to flight we should revisit the earlier decision.[4]

3 Berger, Brian, "Report: NASA should use Shuttle to service Hubble," *Space.com*, December 8th, 2004, http://www.space.com/584-report-nasa-space-shuttle-service-hubble.html

4 Whitesides, George and Barnhard, Gary, "NASA Chief Mike Griffin and Hubble," *Space.com*, April 27th, 2005, http://www.space.com/1009-nasa-chief-michael-griffin-hubble.html

They lobbied him to reverse the decision, and in February of 2008, NASA announced that the mission would proceed in the coming August. It was a rare instance where the astronauts explicitly discussed the risk and reward:

[I]t's a death-defying mission. If something goes wrong with the shuttle [sic] on a routine trip to the space station, astronauts can stay there for up to two months with plenty of food, water and oxygen while waiting for a rescue mission.

On the upcoming mission to Hubble, which orbits 360 miles above Earth, if there's a problem, NASA will have just days to launch a second shuttle [sic] to rescue the stranded astronauts.

Despite the risk, those who volunteered for the mission say they have no second thoughts. "I have thought a lot about that after *Columbia*," said Commander Scott Altman. "This mission is as safe as we can make it and the risk is appropriate for the reward."

Grunsfeld says: "It is about science, it is about inspiration, it is about discovery, it is about all the kids who will look at the Hubble images and dream."[5]

Now here's the irony. As with the decision to remove commercial payloads from the Shuttle, the original decision to cancel was the right one, for the wrong reason. As noted space historian and analyst Jim Oberg observed at the time, the real issue was not crew safety but the risk of losing the orbiter itself and opportunity costs of not doing an ISS mission with that flight.[6] Space journalist and analyst Jeff Foust wrote an article about alternate uses for

5 Champion, Sam, McHugh, Rich and Brady, Jonann, "Astronauts Fight For Hubble Mission And Win," *ABC News*, February 8th, 2008, http://abcnews.go.com/GMA/Story?id=4262727

6 Oberg, James, "Hubble debate a lot of sound and fury," *NBC News*, March 22nd, 2004. http://www.msnbc.msn.com/id/4580820/ns/technology_and_science-space/t/hubble-debatea-lot-sound-fury

the money that the mission would cost (anywhere from a couple hundred million to a billion dollars, depending on how one does the accounting), such as ground-based observatories, or newer-tech space telescopes:

> Here, then, is the question that NASA and the astronomical community must answer: is it worth the cost and the risk to service Hubble again, extending its life for perhaps three to five years, or are there better ways to spend the money and get good science in return? For the cost of a shuttle [sic] servicing mission—say, $500 million—one could get two Medium-class Explorer (MIDEX) missions, currently capped at $170M each. There would still be enough money left over for NASA to build a world-class ground-based observatory (or two) that, unlike Hubble or the Explorer spacecraft, could last for decades. NASA could also "buy in" to existing telescopes, paying for instruments and/or operating costs in exchange for a share of telescope time, which the agency could then make available to astronomers. These alternatives could, combined, do much of what Hubble could do, and provide the public with all the pretty pictures it could want.[7]

In fact, O'Keefe even discussed these issues in his decision, but the prevailing story line was about crew safety. It ended up being a case in which the public's irrational devotion to a telescope overcame NASA's (and Congress's) irrational obsession with avoidance of astronaut deaths.

---

7 Foust, Jeff, "Life after Hubble" *The Space Review*, February 2nd, 2004, http://www.thespacereview.com/article/95/2

Why We Value The Hubble Space Telescope

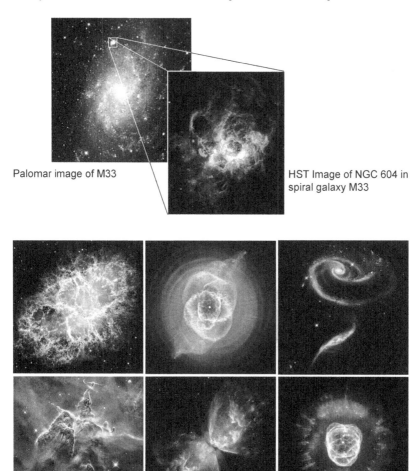

Palomar image of M33

HST Image of NGC 604 in spiral galaxy M33

Just a few of the many spectacular and scientifically valuable images returned by the Hubble Space Telescope (descriptions and credits in "Illustration Credits")

Chapter 8

Safety For NASA In LEO
Going Forward

Starting in the spring of 2011, a battle raged over the best way to procure the new systems needed to eliminate our dependence on the Russians for ISS access and lifeboat services. Over two years later, at the time of this book's publication, it remains unresolved. While most have finally accepted that it is going to be done commercially, there is still a dispute between keepers of the traditional NASA approach, in which the agency issues cost-plus contracts under the Federal Acquisition Regulations (FAR) to selected contractors for the development of hardware that NASA would operate (Apollo and Shuttle models), and a new approach in which services are procured from private providers on a fixed-price-milestone basis (the Commercial Orbital Transportation Services, or COTS) program. This has been simmering over many months, and it has largely been driven by the issue of safety. But cost has been a consideration as well.

COTS had been performed under a type of contract called a Space Act Agreement (SAA). Rather than the traditional approach in which NASA reimbursed a contractor for their costs and added some additional profit, it was a public/private partnership in which the company invested its own money, and was compensated for specified milestones on a fixed-price basis. In other words, there was no payment without the achieved milestone, and if it cost more than the payment, the company ate the difference. It had been very successful in terms of getting the needed services at costs far below what would have been required had NASA developed the capability under a traditional cost-plus contract. The successful test flight of the Dragon spacecraft to the ISS in May of 2012 was in fact a resounding validation of the concept.

Initially, the follow-on program—to do for crew delivery (to and from) the ISS what COTS had done for cargo—was performed on similar Space Act Agreements. But in the summer of 2011, NASA announced a change in plans. For reasons that many viewed as legally arcane, NASA said that in examining the contract language, they had discovered that an SAA couldn't give them the control over the contractor necessary to ensure crew safety.

NASA's original intent, according to Brent Jett, a former astronaut serving as deputy program manager for NASA's commercial crew program, was to use an SAA again for the Integrated Design phase. "As the team dug a little bit further into the Space Act Agreement, we did find several key limitations," he said. The biggest one, he said, is that NASA cannot mandate requirements under an SAA, including for crew safety, but only provide them as a reference for industry. "Even if industry chose to design to those requirements, NASA is not allowed to tie any of the milestones in an SAA to compliance with those requirements," he said. "That means NASA cannot accept the verification of those requirements and certify the system the way we need to for commercial crew under a Space Act Agreement."

Jett noted that, under COTS, NASA was able to exploit something of a loophole in those rules, which allow the agency to levy safety requirements when a NASA facility—the ISS—was involved. NASA could do the same for CCDev, but only for operations at the ISS. "We would not be able to levy any requirements concerning ascent, entry" or any other portions of the flight not directly dealing with approaching and docking with the ISS.[1]

Supporters of commercial crew were up in arms, warning that this would dramatically increase program costs, and likely delay the schedule. But NASA was seemingly firm in its belief

1 Foust, Jeff, "Could a contracting change jeopardize commercial crew?" *NewSpace Journal*, July 21st, 2011, http://www.newspacejournal.com/2011/07/21/could-a-contracting-change-jeopardize-commercial-crew/

that a reversion to a traditional FAR contract was necessary. Unfortunately for the agency, it ran into a roadblock on Capitol Hill, in the form of a tight-fisted Congress, at least when it came to programs that threatened their favorite NASA activity—job preservation in the states and districts of the congresspeople on the relevant committees. NASA had requested $850M for the Commercial Crew Program (CCP) for Fiscal Year 2012, claiming that it was necessary to support its planned FAR contracts. Congress responded with a budget of less than half that—$406M. In December of 2011, NASA reversed course, and announced that it was reverting back to SAAs, at least for the next phase of the program:

> "In a dynamic budget environment, it makes it tough for us to deal with that budget fluctuation" when using fixed-price contracts, [Bill]Gerstenmaier [NASA's Associate Administrator for Human Exploration and Operations] said. "If we don't get the funds that we anticipated, it makes it tough for us to negotiate the contract and inefficiency in renegotiating that contract." Going back to SAAs, he said, "allows us to make significant progress during this period and continue on the way to eventually getting that commercial crew capability for the ISS."[2]

While those supporting the commercial crew program cheered the news, it set off a new round of concerns about "safety" from the pinch-penny solons on the Hill. For example, on February 29th, 2012, Congressman Pete Olson (R-Texas), with several co-signers from the House Subcommittee on Space, sent a letter to presidential science adviser John Holdren to express his concern. Olson is quoted in the Associated Press release:

---

2 Foust, Jeff, "An about face for commercial crew," The *Space Review*, December 19th, 2011,

"Proposed agreements between NASA and commercial crew entities fail to provide adequate authority for safety oversight and guidance needed by NASA to ensure the best safety for astronauts operating independently developed crew vehicles," Olson said. "*Safety is the most critical component of human space exploration. This is a no brainer for NASA.*" [Emphasis added][3]

It would seem that everything is a "no brainer" when it comes to the topic of space safety, particularly on the Hill. The debate was continuing almost literally as I wrote these words in the fall of 2012. On September 14th, 2012, the House Science, Space and Technology Committee held a hearing at which Gerstenmaier was called again to testify on the issue. He declared to the committee that "NASA is committed to ensuring that the requirements, standards, and processes for [c]rew [t]ransportation [s]ervices certification for all commercial missions are held to the same or equivalent safety standards as [g]overnment human spaceflight systems."[4] The committee chairman, Ralph Hall (R-Texas) wasn't buying it:

It's hard for me to understand why NASA is proceeding this way. Will this result in systems that are safe for our American and international partner astronauts? How will NASA know if they don't have the insight? And perhaps more importantly to those of us in Congress who are asked to fund this, how and when will NASA know if it is getting what it needs and if these systems will be safe enough? Redesigns will be costly and time consuming if

---

3 Press release, office of representative Pete Olson, February 29th, 2012, http://olson. house.gov/index.cfm?sectionid=129&sectiontree=21,129&itemid=945

4 "Witnesses Say NASA Must Have Expanded Role in Ensuring Astronaut Safety as Commercial Spaceflight Capabilities Develop," press release, U.S. House Science, Space and Technology Committee, September 14th, 2012, http://science.house.gov/ press-release/witnesses-say-nasa-must-have-expanded-role-ensuring-astronaut- safety-commercial

important technical or safety requirements were not addressed up front.[5]

Note the use of the phrase "safe enough." Vice Admiral Joseph Dyer, USN (retired), chairman of the Aerospace Safety Advisory Panel (ASAP), an organization established after the Apollo 1 fire in the sixties to provide guidance on safety in human spaceflight (when no one in the nation was doing human spaceflight except NASA), expressed similar concerns and phraseology in his own invited testimony:

> Admiral Dyer told Committee Members that NASA's, "current acquisition approach—commercial transportation system development that is funded under a space act agreement concurrent with certification that is funded under a federal acquisition regulation-based contract—is complex and unique. In our opinion, this approach is a workaround for the requirements and communications challenges implicit to the space act agreements."

> Further Admiral Dyer said that NASA "unquestionably face[s] a number of challenges in reaching the point where these [launch] systems can be confidently certified as being '*safe enough*' for the astronauts that rely on this process to *ensure their safety*." Adding that at this point in time designs "are maturing before requirements," and "government and industry have not yet agreed on how winning designs will be accepted and certified. We worry that the cart is ahead of the horse." [Emphasis added][6]

Now, in fairness, in his role as chairman of the ASAP, Admiral Dyer's sole concern is safety. It is not up to him to determine what

---

5 Ibid.

6 Ibid.

is "safe enough," but rather to ensure that however that standard is defined, it is met. But Congressman Hall and his colleagues should be taking a broader view. It is up to them to determine how safe is "safe enough," and how important the missions are for which the astronauts are risking their lives. Yet they never do so, and they continue to take the view that NASA is an expert on space safety, despite the fact that NASA's vaunted "safety" processes killed fourteen astronauts in the Shuttle program.

This perception that NASA is the authority on safety persists from the days of Apollo. For many, NASA has come to represent the pinnacle, the very best in America, if not humanity itself, and tales of cost overruns, mismanagement and inefficiencies won't budge that idealization of the agency.[7] Even the *Challenger* and *Columbia* accidents, with revelations of launch fever and ignored warnings didn't affect the pedestal upon which NASA was put.

Because NASA was held in such high esteem by the public, NASA's management problems—the root cause of both the *Challenger* and *Columbia* accidents—went overlooked. Instead, those tragedies only bolstered the perception that spaceflight itself is difficult. And not just difficult, but exceedingly difficult—so difficult that only NASA can do it. The corollary is that it is risky; so risky that safety can never be assured—because even NASA couldn't do it.

The mind set that only NASA can manage spaceflight to the degree of safety being demanded leaves out a very important driving factor that pertains to the private sector, but not to a government agency such as NASA. And that is, it's in the highest interest of a private company to make their vehicles "safe enough" because, unlike NASA, they won't receive increased budgets when people die in their vehicles. Instead, they will most likely suffer in the market place. It isn't good business to kill your customers.[8]

---

7 Many people, especially conservatives, have a similar attitude toward the military.

8 This is not to imply that individual NASA and contractor employees don't care about losing their friends and colleagues—it's a groupthink problem, not callousness on the part of any team member.

Along those lines, even the Columbia Accident Investigation Board (CAIB) came to recognize the deleterious effects of its 2003 recommendation for crew/cargo separation and its subsequent misinterpretation. When opponents of commercial crew attacked the program by citing the CAIB report to make the claim that NASA was the unique authority on space safety—which was in fact contrary to the findings of the CAIB report (in which NASA's safety policies were actually harshly criticized)—the CAIB attempted to correct this in a letter to Senator Barbara Mikulski (D-Maryland), Chair of the Senate Appropriations Committee for NASA, in the summer of 2010. The CAIB also noted that those who say that only NASA can ensure crew safety are ... mistaken:

[T]he CAIB recommended that future launch systems should "separate crew from cargo" as much as possible. This statement is sometimes taken out of context. What it does mean is that human lives should not be risked on flights that can be performed without people; the new plan to procure separate crew and cargo transportation services clearly is consistent with the CAIB's recommendation. But the recommendation does not disallow the use of a cargo launch system to also fly, on separate missions, astronaut flights. Indeed, the fact that Atlas V and Delta IV are flying satellites right now, including extremely high-value satellites, has helped to prove out their reliability. And the many satellite and cargo missions that Falcon 9 is planned to fly will also produce the same beneficial result.

Third, it has been suggested by some that only a NASA-led effort can provide the safety assurance required to commit to launching government astronauts into space. We must note that much of the CAIB report *was an indictment of NASA's safety culture, not a defense of its uniqueness*. The report (p. 97) notes that "at NASA's urging, the nation committed to build an amazing, if compromised, vehicle called the Space Shuttle. When the agency did this, it accepted the bargain to operate and maintain the vehicle in the safest possible way." The report then adds, "The Board is not convinced that NASA has completely lived up to

the bargain." We commend the efforts of former Administrator
Griffin and current Administrator Bolden and their associates to
address the safety issues raised in the CAIB report and to do
everything in their power to avoid the organizational failures of
the past that led to two tragic Shuttle accidents. However, one
might argue that the similarities in the organizational cause of
both *Challenger* and *Columbia* suggest that it is very difficult for
a single organization to develop, oversee and regulate such a
complex human-rated spacecraft for an extended period of time.
In any event, the operational experience with the Shuttle does
not preclude others from successfully creating a human space
flight capability—as has been demonstrated by the Russians
and Chinese. *We see no reason why a well-crafted NASA-industry
partnership cannot match, or perhaps exceed, past performance in
ensuring astronaut safety.* [Emphasis added] [9]

"[A]n *indictment* of NASA's safety culture, not a defense of
its uniqueness." That should have left a mark. But unfortunately,
much of the mewling about "safety" by Congress is actually a
rationale to continue to prop up the old ways of doing business that
have kept us from making much progress in space. It continues to
ensure that money flows to the right states and districts—a system
that is threatened by any sort of competition.

Also note the glaring inconsistency of declaring commercial
providers "unsafe" because NASA doesn't have sufficient control
over them, when in fact NASA flies its astronauts on Soyuz
launchers and capsules over which they not only have no control,
but have nowhere near the level of technical transparency that
they will get from the commercial providers. Many opponents of
commercial crew have said that a new commercial rocket must
have over a dozen successes before it can be trusted to carry crew.
Yet the Air Force is requiring only three successful launches before
allowing billion-dollar satellites to be carried. NASA had planned
to fly crew on Ares I after only a single test flight; and the very first

9 Letter from Columbia Accident Investigation Board to Senator Barbara Mikulski,
July 12th, 2010, http://www.spaceref.com/news/viewsr.html?pid=34471

flight of the Shuttle—a brand new, extremely complex system—occurred with two men aboard.

If, as Admiral Dyer is concerned, the "designs are maturing before [the] requirements," that is because we have never had an honest national debate about what the requirements are. The providers have been working to the specifications of current NASA human-ratings documents (even though NASA itself has *never* procured or operated a vehicle that meets them), and it is unclear what more they should be doing in that regard, or even if in doing that they are doing the right thing.

So how "safe" is "safe enough"? To answer that question demands that we define the value of what we are trying to accomplish in space, and that necessitates a conversation that we haven't had since the early 1960s.

Over the past decade I've often written that the reason the American public was so upset by the loss of the crew of *Columbia* in 2003 was because they were perceived to have died for something relatively trivial (e.g., a well-publicized part of their mission was performing childrens' science-fair experiments). For that mission, a one-in-a-hundred chance of dying seemed intuitively too high. But suppose instead that they had been on a mission to Mars, or on one to divert an asteroid that threatened our civilization? For the former, would a one-in-ten chance of death be too high and for the latter, how about a fifty percent chance of survival? Would that be "safe enough"?

During WW II, a four percent loss rate of American air crews (one in twenty five) was considered "acceptable." The men who flew Jimmy Doolittle's historic raid on Tokyo, including Colonel Doolittle himself, knew that they had a very high probability of being shot down or captured or killed by people who would take great joy in their suffering. They were all volunteers. They obviously considered the mission worth the risk, even though the primary result of the raid would be only a psychological blow to the morale of the Japanese, not accomplishing a strategic objective in the war.

But for real context here's another astonishing example of which I'd been unaware until researching this book. In transitioning from propellers to jets, the U.S. Navy suffered a horrific number of *non-combat* losses over the years:

> [T]actical jet aircraft design and technology presented Navy aircrews, maintenance personnel, and leaders with several major challenges that were in fact not substantially overcome until the introduction of the F/A-18 Hornet in 1983. These challenges included such technical problems as engine reliability and response times, swept-wing flight characteristics, and man/ machine interface issues. The Air Force also encountered these challenges, but the Navy's operating environment and, indeed, its organizational culture kept it from achieving a fully successful transition until well after the Air Force did.

> Between 1949, the year jets started showing up in the fleet in numbers, and 1988, the year their combined mishap rate finally got down to Air Force levels, the Navy and Marine Corps lost almost *twelve thousand airplanes of all types (helicopters, trainers, and patrol planes, in addition to jets) and over 8,500 aircrew*, in no small part as a result of these issues. [Emphasis added][10]

That's the sort of thing that a nation does when something is important.

Wolfe noted that being an astronaut was safer than being a test pilot. But apparently being a test pilot wasn't much more hazardous, if at all, than simply being a Navy or Marine aviator, or even deck crew. So in light of all that, what should be considered acceptable for spaceflight?

We've already seen the bloody history of exploration, research, technological development, and settlement. Would those frontiers

---

10 Rubel, Robert, "The U.S. Navy's Transition To Jets," *Naval War College Review*, Spring 2010   http://www.usnwc.edu/getattachment/76679e75-3a49-4bf5-854a-b0696e575e0a/The-U-S—Navy-s-Transition-to-Jets

have been conquered with the attitude that they must have been opened with three nines (or more) of safety? Since there is no absolute safety this side of the grave, the only way to absolutely ensure it (at least as far as spaceflight goes), is to never send anyone into space. As American author and professor John A. Shedd once wrote (since memorialized on countless inspirational office posters), "A ship in a harbor is safe[11], but that's not what ships are built for."[12]

In historical terms, the hazards of opening up a new frontier remain with us. Yet societal expectation for safety in all activities has increased. In an age in which people have developed an unrealistic expectation of absolute safety in all activities (and expect government to provide it), we are also attempting to open up space, the harshest and most challenging of frontiers. The conundrum is that these two goals are mutually exclusive, at least at current technology levels (and perhaps at any).

Obviously, in the twenty-first century, we don't value life as cheaply as we have historically, but somehow, modern society has seemingly placed almost infinite value on a human life, at least when it comes to space. If that condition remains, the pattern of the past half century will continue, and very few people will leave the planet—at least Americans.[13]

Are we serious about opening this frontier? If so, we should show it by shedding our irrational aversion to risk. And that starts by declaring forthrightly what we are trying to accomplish by sending people into space as a nation, and deciding just what that goal is worth, in both money and lives (and yes, sorry, there is a trade-off between those two things in space activities, just as there is in any other human endeavor).

To get back to the bizarre (at least that's how it would appear to a Martian) behavior with respect to ISS: what is it worth? Of what value is it to have people aboard? We have spent about a

---

11 Except during a hurricane, when they are actually safer at sea if they can outrun it.

12 Salt from My Attic (1928), The Mosher Press, Portland, Maine; cited in *The Yale Book of Quotations* (2006) ed. Fred R. Shapiro, p. 705

13 China, Russia, Japan and even India may not share our squeamishness.

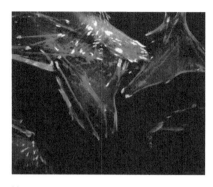

Human monocyte immune cells being researched in ESA's Kubik incubator aboard the ISS

hundred billion dollars on it over almost three decades and we are continuing to spend two or three billion a year on it (depending on how one keeps the books). For that, if the purpose is research, we are getting about one person-year of such, because simply maintaining the facility takes so much crew time that on average, only one person is doing actual research at any given time. That would imply that we think that a person-year of orbital research is worth two or three gigabucks.

What is the constraint on ISS crew size? For now, not volume, though the life-support system may be near its limits (the U.S. segment can supposedly support four people, and the Russian segment three)—I don't know how many ultimately it could handle, but we know that there is currently not a larger crew because of NASA's lifeboat requirement—there has to be a three-person Soyuz for each three people on the station. If what they were doing was really important, they'd do what they do at *Amundsen-Scott*, and live without.

After all, as suggested earlier, just adding two researchers would immediately triple the productivity of the facility. In fact, because the ISS has recently been unable to average more than twenty-seven researcher-hours per week[14], adding one person for a forty-hour week would increase it by two and a half times, and adding a second would increase it by a factor of four. If what we're getting from the ISS in terms of research is really worth three billion dollars a year, then quadrupling it would be, at least in theory, a huge value.

---

14 Smith, Marcia, "ISS Crew Struggling to Find Time to do Research," *Space Policy Online*, November 15th, 2012, http://www.spacepolicyonline.com/news/iss-crew-struggling-to-find-time-to-do-research

In fact, one could go further, and make the case that in our insistence on not risking astronauts' lives, we are causing numerous deaths on the ground. In a post at his *Selenian Boondocks* blog on March 3rd, 2013, space entrepreneur Jon Goff, CEO of Altius Space Machines, noted the very real, but unrecognized cost of the delays of research at the ISS:

One of the most promising applications I've seen for microgravity research on the station is the development of vaccines. Apparently some infectious diseases (I think mostly bacterial ones) behave very different in microgravity—they grow much faster. This increase in virulence combined with turning off some of the confounding factors supposedly enables researchers to more quickly isolate the cell receptors, genes and such that govern the spread of the disease, allowing researchers to craft vaccines that have fewer negative side effects, are more effective, and in theory can make it through clinical testing and to market faster than terrestrial-developed counterparts. At least that's the theory as I understand it, in semi-layman's terms. Two specific diseases are currently being worked on by NASA and commercial firms like Astrogenetix are Salmonella and MRSA (Methicillin-Resistant Staphylococcus Aureus). The theory is that a[n] MRSA vaccine developed on the station will be more effective than a terrestrial version, and will have fewer negative side effects.

So here's [a] thought experiment. Right now, according to a little googling, MRSA kills about 19,000 Americans per year. As I understand it, there are a few terrestrially developed antibiotics and vaccines in the works, but say that a microgravity developed MRSA vaccine was effective 10% more often (i.e. that if say the terrestrial versions could save someone's life 50% of the time, the microgravity-developed vaccine could save someone's life 60% of the time). That would equate to ~1900 lives saved per year, 158 lives saved per month, or approximately 5 lives saved per day. And mind you, those numbers are *only for American lives saved*.

(Quote continues)

Right now the development of vaccines like this are highly dependent on the frequency of up and downmass opportunities on the space station as well as on the crew time available for doing research. From conversations I've had with CASIS, the ISS National Lab, and some others at NASA over the past few weeks, those two challenges (delivery/return frequency and crew research availability) are by far the two biggest challenges to effective use of the ISS. While there are several potential solutions to these problems—and in fact, I'm working on some really intriguing ones on the crew research availability side at Altius at the moment—one of the simplest ways to help improve the situation for both of these problems would be for Commercial Crew to enter operational services quicker.

Right now other than very tiny payloads on Soyuz, Dragon is the only way of getting payloads back from the station, and even when Elon's team gets up to full speed, that's only three opportunities per year. Any of the commercial crew vehicles being developed would add substantially not just to total downmass "tonnage" but more importantly to the *frequency* of downmass opportunities, increasing that number to potentially 5-6 times per year.

Additionally once commercial crew vehicles are flying, their lifeboat capability (and I agree with Rand's take on how necessary that really is) will enable adding an extra crew-person to the ISS, bringing it to a total of seven crewmembers, with four of them on the US side. Right now between the three crewmembers on the ISS, we're only getting about 1800-1900 man-hours of research work done per year on the station, with an average of about 35 hours per week total between the three of them. Just adding an additional crew member on the USOS side would likely double that number, potentially doubling the ROI for the station.

Between these two changes enabled by getting Commercial Crew into operations, experiments like the MRSA vaccine

(Quote continues)

development process can proceed much quicker. As Tom Pickens of Astrogenetix explained at a Space Angels Network event I was presenting at in Houston a bit over a year ago, their development process depends on the ability to do 5-6 launch/process/return iteration cycles during the development of a given vaccine. Adding additional flight opportunities, and making sure that the experiment gets processed while the delivery vehicle is on station so it can make it back on the same vehicle ("Sortie Science" as the National Lab folks are calling it), can both greatly shorten the amount of time it takes to get the vaccine developed and into clinical testing.

While shortening the development cycle has serious positive commercial profitability benefits (a vaccine or design that isn't completed is like a non-interest bearing checking account with a very high monthly fee that you only get profit from once the product actually hits the market), it has a dramatic value in saved lives in this particular case. Put simply, every day a vaccine like this gets to market sooner means a certain number of people who aren't going to die painfully and prematurely. In the particular case of a 10% better MRSA vaccine, we're talking about saving an extra 5 American citizens per day sooner that you get the MRSA vaccine to market.

So what does this have to do with space safety? Pretty simple. If NASA isn't blowing smoke about the benefits of microgravity research for developing vaccines (and I for one believe them in this case), the delays in Commercial Crew availability due to added safety requirements come with an impressive cost in human lives. Adding an extra year to bump the theoretical reliability of commercial crew from 99% to 99.5% for instance just potentially cost you almost 2000 American lives, just from this one vaccine alone. These are lives that could've been saved by allowing a faster, more streamlined commercial crew development process. And by not starving it for funds to pay for heavy-lift rockets without destinations.

(Quote continues)

> Think about that. Just shaving 36 hours off of the availability date of commercial crew could potentially save more lives than would be lost in the worst case Commercial Crew crash. Even if expediting the process, dropping many of the NASA Human Rating requirements, dropping some of the abort tests, and sticking with Space Act Agreements instead of FAR Contracts really meant a massive decrease in actual safety (I don't think it would) to say a 5% chance of losing a crew on a given flight, over the course of the ISS's life you would have saved hundreds of times more US lives by taking that course than you would potentially risk in astronaut lives. [Emphasis in original][15]

One of the problems is that loss of crew is very publicly visible (as we saw with Apollo 1, *Challenger* and *Columbia*), while the people who die from failure to develop or approve drugs are anonymous and unknown to all except those closest to them, and their deaths aren't understood to be a result of flawed government policy. This is the same problem that the Food and Drug Administration (FDA) has. So it often ends up inhibiting innovation, destroying jobs, and as a result killing people lest it be blamed for letting people die through under-regulation. It gets back to the asymmetry described previously about fear of testifying about how the absence of an abort system killed crew, but not about how its presence might have done so.

None of this, of course, is to say that they couldn't be continuing to improve the safety, and develop a larger life boat eventually (the Dragon is probably very close to being able to serve as one now, since it doesn't need a launch-abort system for that role—it only needs a new mating adaptor that allows it to dock to or depart from an unmanned or unpowered station), but their unwillingness to risk crew now is indicative of how unimportant they think whatever science being done on the station really is.

---

15 Goff, Jon, "Crew Safety Perspective," *Selenian Boondocks*, March 3rd, 2013, http://selenianboondocks.com/2013/03/crew-safety-perspective

What is nuclear non-proliferation worth to us? This shouldn't be an issue of civil space policy, but it is. There is a U.S. law called the Iran/North-Korea/Syria Non-Proliferation Act (INKSNA), which states that we will not trade with any nation that supports any of those countries in the development of nuclear weapons and delivery systems. Russia has been doing both for years, and in order for us to continue to utilize their services for ISS access and lifeboats, Congress has to continually waive the law, essentially rendering it toothless with respect to one of the most significant violators of it (in January of 2013, they waived it out to 2020).

After the maiden flight of the Falcon 9/Dragon in 2010, SpaceX founder and CEO, Elon Musk, said (accurately), that if someone had been aboard, "they would have had a very nice ride."

If (as earlier discussed) we were to start using Falcon-9/Dragon sooner, even without its abort system, we could stop depending on the Russians, and stop shipping money to a nation that is indifferent to our security, if not outright hostile to it. Why don't we? Because we don't want to risk the lives of an astronaut crew, even though the Falcon-9/Dragon is probably as, or more, reliable at this point than anything we flew in the 1960s. The same thing applies for the Atlas launcher and the Boeing Crew Space Transportation (CST) capsule, another proposed Commercial Crew solution.

I think that it's "safe enough" right now to end our dependence on the Russians. Despite the stated desire for three nines of safety, I'd bet that most people in the astronaut office would agree, and if there are some who don't, no one held a gun to their heads to be an astronaut. In our unwillingness to do this, we are saying that the life of an astronaut crew is more valuable than preventing Iran from getting nuclear weapons, or to be more precise, we don't think that non-proliferation is worth risking their lives. I don't think that's the case, and I'd guess that most astronauts don't, either, but in its continuing hyperconcern about safety, that is exactly the message that we are getting from Congress. Putting this in perspective, we see many men and women willing to risk their lives for national security every day, in Afghanistan and other dangerous places. If I were an astronaut, I'd be insulted if Congresspeople didn't think that I'd also be willing to risk my own.

How important is it to send government employees to other planets? Well, seeing how little progress we've made in that regard in the four decades (almost exactly as this book is being completed) since Gene Cernan last stepped on the moon in December of 1972, one is tempted to say, not very. However, Congress continues to pretend that it is, by insisting on wasting tens of billions of taxpayer dollars on a "big monster rocket" (to use Senator Bill Nelson's phraseology) that NASA's own internal studies indicate that it doesn't need to get to other planets, and is absorbing all of the budget needed to build things that it actually does need to do so, such as landers and orbital propellant storage facilities. But how much should we be willing to risk the life of an astronaut in such an endeavor? What is it worth? As science-fiction author Jerry Pournelle once commented,[16] "Astronauts apparently aren't national heroes, but rather national treasures, too precious to be risked on something as hazardous as spaceflight."

I've written that today's NASA couldn't do Apollo 8. If that's true, then we have no hope of sending humans to Mars, because the risks involved in Apollo 8 pale in comparison to such an expedition, and they make the risk of ascending into low earth orbit positively trivial. (Again, as a reminder, this was why Ares I made no sense if the purpose was to spend many billions to ensure safety on the safest phase of the mission). If we seriously want to send Americans to Mars, then we should be taking a cue from

---

16 Personal email, Henry Spencer. He writes: "In the 'Near-future Space' panel at the 1992 World Science Fiction Convention in Orlando—Jerry said he'd told Vice President Dan Quayle that the orbital Delta Clipper test vehicles should be flown not by astronauts, who are 'national treasures,' but by test pilots, because if one dies you name a street after him at Edwards and carry on. Astronaut John Young, who was on the panel, was visibly squirming."

Magellan, and send an armada, with the implicit assumption that we're going to lose some ships, and lose some explorers or settlers. Bob Zubrin, veteran space engineer and head of the Mars Society made a similar point early in 2012 in *Reason* magazine:

[A]m I saying that we should just bull ahead, regardless of the risk? No. What I am saying is that in space exploration, the top priority must not be human safety, but *mission success*. These sound like the same thing, but they are not. Let me explain the difference by means of an example.

Imagine you are the manager of a Mars robotic-rover program. You have a fixed budget and two options for how to spend it. The first option is to spend half the money on development and testing, the rest on manufacturing and flight operations. If you take this choice, you get two rovers, each with a 90 percent chance of success. The other option is to spend three-quarters of the budget on development and testing, leaving a quarter for the actual mission. If you do it this way, you get just one rover, but it has a success probability of 95 percent. Which option should you choose?

The right answer is to go for two rovers, because if you do it that way, you will have a 99 percent probability of succeeding with at least one of the vehicles and an 81 percent probability of getting two successful rovers—an outcome that is not even possible with the other approach. This being a robotic mission, with no lives at stake, that's all clear enough. But if we were talking about a human mission, what would the right choice be? The correct answer would be the same, because … the first obligation must be to *get the job done.*

Of course, if the choice were between two missions that each had just a 10 percent success probability and one with a 90 percent chance, the correct answer would be different. The point is that there is a methodology, well established in other fields, that can help assess the rationality of risk reduction expenditures

in the human spaceflight program. If NASA disagrees with the
suggested assignment of $50 million for the life of an astronaut,
it should come up with its own figure, substantiate it, and then
subject its proposed plan of action to a quantitative cost-benefit
analysis based on that assessment. But it needs to be a finite
number, for to set an infinite value on the life of an astronaut is
to set both the goals of the space exploration effort and the needs
of the rest of humanity at naught.[17] [Emphasis added]

Unfortunately, when it comes to space, Congress has been
pretty much indifferent to missions, or *mission success*, or *"getting
the job done."* Its focus remains on "safety," and in this regard,
price is no object. In fact, if one really believes that the reason
for Ares/Orion was safety, and the program was expected to
cost several tens of billions, and it would fly (perhaps) a dozen
astronauts per year, then rather than the suggested value of fifty
million dollars for the life of an astronaut, NASA was implicitly
pricing an astronaut's life to be in the range of a billion dollars.

As another example, if it were really important to get someone
to Mars, we'd be considering one-way trips, which cost much
less, and for which there would be no shortage of volunteers.[18]
It wouldn't have to be a suicide mission—one could take along
equipment to grow food, and live off the land. But it would be very
high risk, and perhaps as high or higher than the early American
settlements, such as Roanoke and Jamestown. But one never hears
serious discussion of such issues, at least in the halls of Congress,
which is a good indication that we are not serious about exploring,
developing, or settling space, and any pretense at seriousness ends
once the sole-source cost-plus contracts have been awarded to the
favored contractors of the big rockets.

---

17 Zubrin, Robert, "How Much Is an Astronauts' Life Worth?", *Reason*, February,
2012, http://reason.com/archives/2012/01/26/how-much-is-an-astronauts-life-worth/1

18 A group in Holland proposed a reality television show based on exactly such a
premise, and in early December of 2012, it announced plans to start sending people
by 2023 as a non-profit institute called Mars One. It seems a more likely scenario
to me than a government-funded expedition, and tens of thousands have already
volunteered.

For these reasons, I personally think it unlikely that the federal government will be sending humans anywhere beyond LEO any time soon. But I do think that there is a reasonable prospect for private actors to do so—Elon Musk has stated multiple times that this is the goal of SpaceX, and why he founded the company. In fact, he recently announced his plans to send 80,000 people to Mars to establish a settlement, within a couple decades, at a cost of half a million per ticket.

On December 6th, 2012, the day before the fortieth anniversary of the launch of Apollo 17, the last NASA manned mission to the moon, a new private venture was announced. Named "Golden Spike" (after the completion ceremony of the first transcontinental railroad), their goal is to use existing launch systems with new hardware for landers and insertion stages to offer two-person trips to the lunar surface for $1.5 billion per flight. They estimate that it will require several billion dollars up front to develop the infrastructure. But they are hoping that by filling an order book, they will be able to raise or borrow the funds. The company claims to already have expressions of interest from sovereign clients—small nations who would like to have an astronaut program on the cheap. It's a similar business approach to that of Bigelow Aerospace with their private orbital facilities. How much safety will the customers demand? Who knows? Surely some will demand more than others, and if the company can't affordably satisfy that demand for enough customers, it won't succeed. But the reliability requirement won't be either arbitrary or infinite, as currently seems to be the case.

Proposed lunar lander for Golden Spike private-sector expeditions to the moon

On February 27th, 2013, an even bolder proposal was described in a press conference at the National Press Club by Dennis Tito, the first man to pay for a flight to orbit with his own money.

Early artist's conception of Inspiration Mars capsule and habitat module passing by Mars on its 500-day journey to fly-by Mars and return to earth

Called "Inspiration Mars," the plan is to send a married couple, past child-bearing age, on a mission to the Red Planet, swoop by within a hundred miles of its surface, and return to earth. The duration will be about seventeen months. The boldest part of the plan is that they will depart earth orbit on January 5th, 2018. That is, less than five years from now. Contrast this with NASA's stated plans of missions to Mars no sooner than the late twenties.

The idea is to take advantage of a planetary alignment that allows a fast trip at relatively low cost in propellant. Such an alignment only occurs a couple times every fifteen years. The one in 2018 occurs coincidentally in the semi-millennial year of Magellan's circumnavigation of the earth, and the fiftieth anniversary of the circumnavigation of the moon by Apollo 8, in December.

So the rocket-scientist-turned-financier, Dennis Tito, who dreamed up and is providing seed funding for the first two years of the mission, sees it as an auspicious date to take a trip around Mars. The next such similar opportunity won't be until 2031, at which point he expects an armada from multiple nations to perform the feat. But as he noted in the press conference, at that time he'll be in his nineties and doesn't want to wait that long.

The Inspiration Mars mission approach is in stark contrast to standard NASA practices. The vehicle will consist of an entry capsule (a SpaceX Dragon was chosen as a reference design) with an additional expandable pressurized module to provide adequate volume to meet both psychological and practical needs for a five-hundred-day journey. In a paper he presented at an Institute of Electrical and Electronics Engineers (IEEE) conference a few days later, Tito proposed a mission architecture that could be performed with a single launch of a Falcon Heavy and indicated that the entire project could be performed for far less than one billion dollars—an amount well within range of private funding.

It will be a spartan trip. As Jane Poynter, co-founder of the environmental engineering firm Paragon Space Development Corp. noted, it will be like a very long road trip in a Winnebago, except you can't get out. To save mass, they'll be recycling water, as NASA has been doing experimentally on the ISS. "They'll have three thousand pounds of dehydrated food, yum," she said. "And it will be rehydrated with water that they drank a couple days before," and a couple days before that, and before that.

A crew of two is the smallest crew that still provides redundancy in the event of the loss of one, and three would introduce potentially risky psychodynamics for a long mission. But even given that, the choice of a married couple isn't just for practical psychological reasons – it is also planned to be symbolic, and fully representative of humanity as the first emissaries to another planet in our solar system. The primary goal in Tito's mind is to inspire young people in a way that NASA hasn't done since the days of Apollo, and for the Marsonauts to be people who both boys and girls can imagine themselves being.

They will be older, and have had their children already (or plan to have none) for two reasons. First, the risk of birth defects in a child conceived after return would be high due to the long radiation exposure. But more importantly for the mission, it would be to prevent the risk of a pregnancy in a weightless, high-radiation environment, that would not only be bad for a developing fetus (assuming that it's even possible to conceive in weightlessness), but impair mission efficiency itself, because the crew are a vital component of the mission. The life-support system will not be automated, but simple and robust and repairable, "like a 1955 Chevy," as Taber MacCallum, Paragon co-founder and chief technical officer for the venture, described it at the press conference. If the crew isn't healthy, they won't be able to keep the ship healthy, either.

In fact, for that reason, this is a mission in which safety really is appropriately the highest priority, because the mission will fail if the crew doesn't return. Tito himself said at the press conference that if he didn't think they had a 99% chance of surviving, he wouldn't approve the flight (though since he has only promised to fund the first two years, someone else may make that decision, and come up with a different number). Because there is no landing on, or even orbit around Mars, the primary challenge is to simply keep the crew alive and healthy. It is an exercise in extended life support in the harshest environment ever. The only "science" to come from this mission will not be planetary science, but to learn about human physiology and its capacity to survive in such an environment, both physically and psychologically. The only real mission-critical events are the departure burn from earth orbit (after which the vehicle is going to Mars, or at least somewhere) and the small correction burns later to ensure a proper Mars passage and later earth-atmospheric entry. Everything else is just keeping crew alive for the duration, with no mission-abort opportunities after the trans-Mars injection.

The biggest challenge, other than psychological, will be radiation. One way to shield against solar events will be to carry the earth-departure stage along for the entire trip, and maneuver it to protect from unidirectional energetic particles. However, for the steady background radiation other solutions will have to be found, including screening for individuals less susceptible to it, medications, and shielding.

MacCallum and Poynter might be the perfect candidates for the crew, given that they will be designing the equipment and are not just co-founders of the company, but founded it while living together (with six others) for two years isolated in the *Biosphere II* closed-environment experiment in Arizona two decades ago;

they were subsequently married. The only issues might be their willingness to go, and her current state of fertility. Jonathan Clark, former NASA flight surgeon (and widower of Laurel Clark, who was killed in the *Columbia* disaster in 2003) stated that the crew won't need to be selected until six to twelve months prior to departure, so in any event it's not an immediate issue.

Some in the audience at the press conference, steeped in decades of the hyperexpensive NASA way, were skeptical of the feasibility. Seth Borenstein of the *Associated Press* led off the questioning with a litany of issues—weren't they going to have a test flight, wasn't this too risky, how could they possibly do it in only five years, what response did they have to all of the unnamed experts with whom he'd consulted who claimed this wasn't feasible? To paraphrase his long-winded rant, "Are you crazy"?

In response, Tito noted (as I would have) that Apollo 8 had no test flight prior to sending humans around the moon, and the very first Shuttle flight had a crew. All the panelists pointed out that yes, it was risky, but that there are some things worth taking risks for, though many in today's America seem to have forgotten that (a familiar theme to anyone who has read this far into this book).

Will he pull it off? Master of Ceremonies and veteran space reporter Miles O'Brien characterized the key mission properties as "simplicity, audacity and liquidity" (the latter referring to the fact that it was funded, for now, and the hardest part of any space mission is always raising the money). The first man to buy a ride into space is tenacious, and he's hired the best in the business. I wouldn't bet against him.

The point is that there are a number of other people with similar goals and financial resources, and one or more of them is likely to achieve them, because they have a much different tolerance for risk than bureaucrats and Congresspeople. The real danger of NASA hypersensitivity to safety is not that it will prevent NASA from sending astronauts into space, but that its overly stringent standards will osmotically bleed over into the regulation of commercial human spaceflight, making it more unaffordable for all, at least in the USA.

And the danger of that occurring is made all the greater by well-meaning recommendations and suggestions from influential players. In early January of 2013, the aforementioned ASAP issued its annual report for 2012, in which it expressed concerns about different safety regimes for commercial versus NASA spaceflight. It seems to be schizophrenic in both those concerns, and its recommendations:

> The ASAP is concerned that some will champion an approach that is a current option contained in the Commercial Crew Integrated Capability (CCiCap) agreement. There is risk this optional, orbital flight-test demonstration with a non-NASA crew could yield two standards of safety—one reflecting NASA requirements, and one with a higher risk set of commercial requirements. It also raises questions of who acts as certification authority and what differentiates public from private accountability. Separating the level of safety demanded in the system from the unique and hard-earned knowledge that NASA possesses introduces new risks and unique challenges to the normal precepts of public safety and mission responsibility. We are concerned that NASA's CCiCap 2014 "Option" prematurely signals tacit acceptance of this commercial requirements approach absent serious consideration by all the stakeholders on whether this higher level of risk is in fact in concert with national objectives.[19]

There's a lot to unpack there. First, let us note that the notion of having multiple standards of safety is apparently a "risk." That is, it is something bad, to be mitigated. Note also the assumption that the commercial standard will be "higher risk."

Let's leave aside their confusing use of the word "risk" to refer to the level of safety, when it generally connotes a programmatic issue (as two examples, schedule risk, or risk that the program will overrun its budget). In this case, in that sense, they are really

---

19 Aerospace Safety Advisory Panel, Annual Report for 2012, Cover Letter, January, 2013

describing the risk that the program will not meet their safety standard (whatever that is).

My first question to them would be: why should NASA requirements be higher (in the sense of more safe) than commercial ones? And why would they think that there will only be one set of commercial requirements? As I've discussed previously, and will expand on in the next chapter, commercial requirements will likely be tailored to the customer and the mission, and will vary among vehicle types and providers. As I've argued here, NASA's should as well. For example, whether performed by NASA or a more appropriate agency, an asteroid diversion mission should allow a higher probability of loss of crew than a routine flight to do ISS research.

Now there are answers to their concern about who acts as the "certification authority." For a flight carrying a NASA astronaut, NASA will be given insight into both the design and operations process to provide them some non-zero level of comfort, though neither the vehicle nor its operations will be "certified" in an FAA-aviation sense. And the differentiation between "public" and "private" accountability will be no different in this case than it currently is between NASA and the Russians who (as noted previously) are much less transparent in that regard than will be the American commercial operators.

It's not clear what the next sentence even means. How does one "separate a level of safety" from "the unique and hard-earned knowledge that NASA possesses"? And what are the "normal precepts of public safety and mission responsibility"? Public safety (that is to say, safety for those not voluntarily involved in the flight) isn't really an issue at all, other than the degree to which it is already protected by the FAA's launch-licensing process (which has nothing to do with NASA). The issue at hand is not public safety, but safety of the spaceflight participants themselves, whether company employees or NASA civil servants or, ultimately, fare-paying passengers. Under current rules, they will be flying on an informed-consent basis.

And note that they apparently didn't read the CAIB's response to earlier foolish plaudits of NASA's safety culture, in which the

CAIB reminded all that their report was an indictment of it, not a "defense of its uniqueness."

This is largely based on the false perception that organizations have knowledge and experience independently of their employees. When I worked in business development for a government space contractor, I was always amused by the standard section we'd always have to put in our proposals to NASA or the Air Force about our company's previous experience and heritage, as though the people who'd worked on those programs in the sixties weren't dead or retired.

Organizations don't have knowledge—individuals do. And to the degree that NASA has any knowledge, it is because it has retained employees who have it. But many of those knowledgeable people have instead gone to work for the commercial companies, so there really is nothing "unique" about NASA. But to the degree that there is, it is primarily that, at least with respect to safety, its procedures have resulted in the loss of fourteen astronauts in flight.

Note also the double standard. When NASA planned to fly crew on the Ares I after a single test flight, there was nary a complaint from the panel. This creates a perception, real or not, that the ASAP is not so much concerned with safety as with being a booster and cheerleader for the government space agency. This perception was only reinforced by its 2009 annual report, when the Ares program was perceived to be in political trouble in the wake of the Augustine Committee report, in which the ASAP warned against "[s]witching from a demonstrated, well-designed, safety-optimised [sic] system to one based on nothing more than unsubstantiated claims ..."

Of course, the Ares I hadn't been demonstrated, and the ASAP repeated the previously noted error of confusing "human rated" components with a human-rated system. In fact, its defense of the Ares I and the Ares I-X test flight in that report was almost laughable. (For me, it was more than "almost"—I literally laughed when I read it.)

On October 28, 2009, NASA's Ares I-X test rocket lifted off from the Kennedy Space Center's (KSC) newly modified Launch Complex 39B for a 2-minute powered flight. The flight test lasted about 6 minutes until splashdown of the booster stage about 150 miles downrange. The flight test will provide NASA with data that will be used to improve the design and safety of the U.S. next generation space flight vehicles that could take humans beyond low-earth orbit (LEO). By flying the vehicle through first stage separation, the test flight provided data regarding the performance and dynamics of the Ares I solid rocket booster in a "single stick" arrangement. The data also allows NASA to establish confidence in the preflight design and analysis techniques. *It should be noted that Time Magazine cited the Ares rocket as the "best invention of 2009."* [Emphasis added][20]

Here's what the report didn't report about that flight test—it didn't test the vehicle that would eventually become the Ares I. It used a four-segment solid-rocket motor, with a dummy segment inserted on top of the four live ones, instead of the five-segment configuration planned for the operational vehicle. It had the same outer mold line as the operational vehicle, so it did test the aerodynamics. But its mass and propulsive properties were quite different, so all they got was validation of their simulation models (not without value, to be sure). They also don't mention that one parachute completely failed, and another partially failed, resulting in the supposedly reusable stage hitting the ocean so hard that it was dented, ruining it for further flights. And it did this for about half a billion dollars—*the same amount of money that SpaceX spent to develop their Merlin rocket engines, both the Falcon 1 and Falcon 9 launchers, their pressurized, returnable Dragon capsule, and all the associated test, manufacturing and launch facilities.*

The last line about that noted expert publication on space technology, *Time* magazine, endorsing the vehicle as the "best invention of 2009," (or any year, for that matter) was the subject of

---

20 2009 ASAP Annual Report, January, 2010, Introductory Remarks Page 2, http://oiir.hq.nasa.gov/asap/documents/2009_ASAP_Annual_Report.pdf

much hilarity in the space community when it occurred, and didn't enhance the credibility the ASAP report, but rather diminished it, and almost had a look of desperation.

The ASAP is going to have to accept the fact that private companies are going to fly their own test crews, to their own planned or expected levels of safety, and that while the panel can certainly offer advice, that is its only role and authority. It should also accept the possibility that this is not necessarily a bad thing. As analyst Clark Lindsey noted at the time at *NewSpace Watch*:

> ASAP expresses concern about the initial testing of the vehicles with the CCP firms' own crews. Under the current commercial space transportation regulatory regime, a company can build a rocket system and fly people to orbit under a FAA license. The company need never have any interaction with NASA whatsoever. Whether that necessarily means it is a less safe transport than one "certified" by NASA is a matter of debate but it is nevertheless quite legal. Simply because a company in CCP accepts NASA funding and technical input and oversight does not mean it gives up the right to fly its own crews. The commercial crews, which no doubt will include former NASA astronauts, are allowed to accept the risks. (The issue of "public safety" enters with the FAA license requirement that there be extremely low risk to third parties during launch and return.)

> The issue for ASAP should be whether those test flights will provide the data needed to tell whether the systems are sufficiently safe for NASA crews. (ASAP might also try to define exactly what level of safety they consider sufficient.) However ... flight data seems not to weigh heavily with the panel.

> If ASAP has its way, a possible scenario would see CCP firm(s) enter the certification phase and spend years there struggling to be approved to fly NASA crews. And in the meantime, the firms will be flying crews and passengers routinely to orbit to a Bigelow station and for other LEO activities. *The absurdity of*

such a scenario would be a logical consequence of the convoluted and contradictory views on astronaut safety held by ASAP as well as by NASA, Congress and the public. [Emphasis added][21]

One would hope that as that scenario actually played out, its absurdity would become increasingly clear to all and result in more sensible policy. As noted previously, there is no intrinsically correct level of risk for either NASA flights or commercial ones. In both cases, the risk will have to be balanced against the payoff. And the commercial industry will have to fight very hard to ensure that a low-risk high-cost NASA standard doesn't become an unaffordable industry one. Just what commercial industry standards should or might be is the subject of the next chapter.

---

21 Lindsey, Clark, "NASA Aerospace Safety Advisory Panel's annual report + some commentary ," January 10th, 2013, *NewSpace Watch,* http://www.newspacewatch. com/articles/nasa-aerospace-safety-advisory-panel039s-annual-report-some-commentary.html

Chapter 9

Safety For Commercial Spaceflight

Take me out to the black
Tell them I ain't comin' back
Burn the land and boil the sea
You can't take the sky from me
— Joss Whedon

While NASA has become extremely risk averse as a result of its history and political environment, the commercial spaceflight industry must be more realistic if it is to thrive. Fortunately, there are some signs that it will be allowed to do so. For instance, the moratorium on FAA regulation of the safety of spaceflight participants first put into place in 2004 was extended in 2012 by Congress until at least October 2015. This was for two reasons, one practical, and the other philosophical.

Practically, we simply don't have enough flight experience with these kinds of vehicles to know what best practices and standards are. In fact, we have almost none at all. The original 2004 act that provided for FAA regulation of passenger safety in commercial human spaceflight[1] had an eight-year moratorium that would have expired in 2012, and the thought was that by that time suborbital providers such as Virgin Galactic and perhaps others would have been flying and such experience would have been developed. But as of this writing, for a number of reasons beyond the scope of this book, there has still been no commercial spaceflight of humans, so it made sense to extend it to late 2015, as congress did in 2012, (and likely beyond). To argue that the

---

1 They have, since their inception in the mid-eighties, always had the ability to regulate the industry with regard to public safety—that is, to protect uninvolved third parties on the ground.

moratorium should be ended now is to argue that it should never have been in place at all.[2]

Beyond that, there are multiple approaches to vehicle design, just as there were in the early days of aviation. Back in the twenties, we hadn't yet figured out how many engines aircraft should have, how many wings, where the engines should be located, whether control surfaces should be in the front or the rear, etc. Had we established standards and certification in the 1920s equivalent to today's aircraft regulations, it's likely that the industry would have been stillborn, or forced into some less-than-optimum standard.

We are at the equivalent state with spaceflight as we were nine decades ago with aircraft. And without saying which is safer, vertical approaches are unlikely to be exactly the same safety level as winged ones, or liquid versus hybrid or solid propulsion. The various near-term planned approaches are summarized in Table 1. Let's look at three suborbital cases to demonstrate the differences in approach, and their different ways of giving those aboard a bad day. This is not to make the case that any approach is safer than any other, but simply to point out that they are different, and that we don't currently have the operational experience to know which is best from a safety standpoint. Moreover, even if one were less "safe" than another, that is not necessarily a reason not to fly it—it may be less expensive, or it may accomplish a goal that another won't. We make such decisions all the time in other transportation choices—all else being equal, a large car is "safer" than a small one (at least in terms of crash resistance, though it may have a higher probability of having an accident), but it costs more. It's all part of life's trade-offs. At this stage of the space-transportation industry, this is something that should be determined by an informed marketplace, not a federal bureaucrat.

So the next few pages should not be viewed as an advertisement for, or against, any of the approaches. But it's useful to consider it the beginning of the "informed" part of the informed-consent regime in which these companies currently operate from a regulatory standpoint.

---

2 Not to imply, of course, that there aren't those who would so argue.

Table 1: There is a Broad Variety of Commercial Approaches to Spaceflight

Company	Armadillo	Blue Origin	Sierra Nevada	SpaceX	Strato-Launch	Virgin Galactic	XCOR
Vehicle	Various	Various	Dream Chaser	Falcon 9/Dragon	TBD	SpaceShip-Two	Lynx
Suborbital	Y	Y	N	N	N	Y	Y
Orbital	N	Maybe	Y	Y	Y	N	N
Passengers	Y	Y	Y	Y	TBD	Y	Y
T/O Mode	Vertical	Vertical	Vertical	Vertical	Horizontal	Horizontal	Horizontal
Landing Mode	Vertical	Vertical	Horizontal	Vertical	TBD	Horizontal	Horizontal
Air Launched	N	N	N	N	Y	Y	N
Expendable	N	N	Y	TBD	TBD	N	N
Reusable	Y	Y	Partially	TBD	Partially	Y	Y
Initial Operations	Spaceport America	Van Horn, Texas	Cape Canaveral	Cape Canaveral	Mojave, CA	Mojave, CA	Mojave, CA Midland, TX
Eventual Operations	TBD	TBD	KSC	KSC + TBD	KSC + TBD	Spaceport America + TBD	Curacao, KSC + TBD

Lynx

XCOR Aerospace's vehicle, called Lynx, is a two-person (pilot and passenger) rocket plane that takes off from a runway, ascends to a hundred kilometers altitude, enters the atmosphere and glides back to the runway from which it took off.

So what could go wrong? In terms of propulsion, the Lynx burns kerosene and liquid-oxygen propellants. The company says that it has never had a hard start in thousands of engine firings, but there could always be a first time. There is a blast shield around each engine and it will be strong enough to minimize fratricidal damage if an engine decides to suddenly disassemble itself in an energetically enthusiastic manner.[3] If there is a total loss of thrust, it would mean simply a mission failure, unless it occurred right after takeoff, in which case it could possibly kill or injure both aboard (as would be the case for an aircraft as well), though this is unlikely, given the small size of the vehicle.[4] Unlike airliners, off-field landings of light aircraft (which the Lynx basically is) are generally survivable. More likely, it would simply result in loss of vehicle. If it has enough altitude, the fact that it utilizes liquid propulsion means that propellant can be dumped in the event of an early return to base, which would result in a normal (i.e., dead stick, unpowered) landing.[5] The landing gear could fail to deploy, or only partially deploy (again, a problem shared with aircraft in general). This is unlikely, because there is a manual, gravity backup in the event of a failure of the powered deployment system. And even then, it wouldn't necessarily result in the death or injury of occupants, but it could.

Another potential problem would be jammed aerosurface controls or some other loss of function. (The vehicle doesn't utilize hydraulics or other assist, let alone fly by wire—it just uses conventional light-aircraft mechanisms). Again, this is unlikely as

---

3 The company has reportedly tested the blast shield with an explosive charge equivalent to a worst-case engine explosion.

4 This is a problem only for fields without a cross runway, which allows an early abort by making a quick turn onto it. Initial test fields, such as Mojave Air and Spaceport, or Midland International Airport, have such runways.

5 XCOR has designed it to be dumped within a minute.

Artist concept of the XCOR Lynx rocket plane

it is a very mature technology, but if it occurred it could result in a loss of vehicle and both aboard. It has small rockets for attitude control while out of the atmosphere. This system should be very reliable, and it has full redundancy, but if it doesn't work, and the entry orientation is wrong, it could result in a loss of vehicle control, and potential for overheating with resulting structural failure, or loss of pilot consciousness. Either of these would be Bad Things.

But the key point, and reason that loss of pilot consciousness would be a big problem is that the pilot is a "single-string" critical system (the XCOR people call it a "meat servo"). That is, there is no redundancy. For security reasons (thank you, Osama bin Laden), the passenger has no controls, so if the pilot can't fly the plane, both aboard die.[6]

This brings us to the next thing that can go wrong: a loss of cabin pressure. XCOR has decided to cover this risk by requiring both pilot and passenger to wear pressure suits, providing "belt and suspenders," though it can potentially detract from the experience for those seeking it[7], and makes it harder for researchers to push buttons or pull levers. Suiting only the pilot would recover the vehicle, but would risk the health and life of the passenger. Other

6 The controls on the right side of the vehicle are removed for passenger flights. They can be installed for pilot-training purposes.

7 Except, of course, for those who view being suited as an intrinsic part of the experience, and there are likely many.

pilot-related risks are the pilot suffering a heart attack or stroke during the flight, or simply making an error (hey, anyone can have a bad day). In any of these scenarios, both pilot and passenger are toast (and if the passenger is a trained pilot, she is particularly frustrated to be auguring in and unable to do anything about it because she's in the wrong seat).

It should be noted, though, that as a last line of defense, Lynx pilot and passengers will have parachutes built into their space suits, though not ejection seats (the company considers them an unnecessary hazard, and designed the outboard wing strakes to allow safe egress, with seat boosters). So in some limited circumstances, one or both may be able to escape a problematic vehicle and live to fly another day.

Now, let's contrast this horizontal takeoff/landing system with another: Virgin Galactic's WhiteKnightTwo (WK2) and SpaceShipTwo (SS2).

SpaceShipTwo

WK2, with SS2 suspended beneath, takes off under jet propulsion from a runway. At altitude SS2 separates, lights its rocket, goes into space, glides back to earth and lands. WK2 flies back and lands under jet power. How does this differ from Lynx?

Well, first of all, it's a lot larger—it can carry six passengers, rather than just one. It has a much larger cabin, and the passengers will be allowed to float around in the several minutes of weightlessness after the engine shuts down. It will have two pilots rather than one. And no pressure suits will be required (they are trusting that the hull won't fail). However, if there is a hull breach in space, the passengers could be injured or killed, and if the crew isn't suited, the ship would go out of control and probably crash. It isn't clear whether or not the crew will be suited against this eventuality, to allow them at least to get the plane back into the atmosphere as quickly as possible, though there will be physical limits to this. Given their unpowered trajectory in essentially a vacuum, they will be hostage to gravity until they enter the atmosphere. If the crew is suited, it could raise the same

Virgin Galactic SpaceShipTwo suspended under first-stage WhiteKnightTwo

psychological issue for the passengers as the earlier example of an aircraft flight crew wearing parachutes.

Another difference is that the vehicle utilizes a hybrid rocket (solid rubber fuel and liquid nitrous oxide for an oxidizer) rather than the all-liquid system that XCOR uses. This was done in the name of safety. The thinking was that, in theory, a hybrid system can't explode because it can't mix the fuel and oxidizer quickly enough. Unfortunately for that argument, several people were killed or injured in the summer of 2007 when a test article exploded in a cold-flow static ground test using just nitrous oxide.

So is an explosion of the propulsion system likely? Probably not, but it's a bad thing if it happens, with potential loss of vehicle and all aboard. Because it's air launched, there is no danger from a failure to light or continue to burn near the ground. The carrier aircraft could have a propulsion failure at takeoff. It is unlikely these days, particularly with four engines, but it can happen. Virgin Galactic claims that SS2 is capable of landing with a full fuel load in the event of an abort. It can dump the oxidizer, but it can't eject the solid rubber grain in the motor. So, in terms of structure, landing gear loads and center of gravity, it's designed to glide and land with that extra weight.

One other key difference is a unique feature of SS2 that its designer, Burt Rutan, invented to ensure that the vehicle always enters the atmosphere in the correct attitude on its return to earth—a reconfigurable wing that turns the ship temporarily into something resembling a badminton shuttlecock, with a preferred orientation in the face of drag. Both the Lynx and SS2 have reaction control systems (RCS) (small rocket motors) to allow maneuverability in pitch, yaw and roll while above the atmosphere. But Rutan's wing eliminates the failure scenario to which the Lynx is susceptible—hitting the upper atmosphere in the wrong orientation. The Lynx relies on its attitude-control system for orientation. However, SS2 will passively align to the proper orientation no matter how far it is from optimal when it hits the upper atmosphere. Of course, it requires that the SS2 wing properly reconfigure for entry, and then restore to its normal mode for landing. While it will be tested thoroughly (and already has been in the atmosphere[8]), a failure of achieving either configuration would make for a bad day. That is because even if it enters properly, if the wing can't be returned to normal, it won't have the proper flying and, more importantly, landing characteristics. I don't know what the terminal velocity would be in that configuration. It may even be possible that the crash would be survivable. But the passengers would, at the least, want their money back.

Like Lynx, a partial or non-deployment of the landing gear would be a problem, but probably not fatal. And like Lynx, pilot error or disabled pilot is possible, but very unlikely with redundant pilots (absent a problem with the aircraft itself).

So that takes care of the two competitors for horizontal operations. How about vertical?

Armadillo

The two vertical companies (so far) who are planning passenger flights are Blue Origin and Armadillo Aerospace. I won't comment on Blue Origin, because due to the very secretive nature of the company, little is publicly known about their vehicle designs or operations. As of the summer of 2013, Armadillo has gone into hibernation for lack of funding. But their most recent concept was a vehicle that takes off vertically, accelerates upward to

---

8 In fact, it was reportedly used to stabilize the vehicle when it seemed to go out of control in a flight test in the fall of 2011.

above the atmosphere, and then enters
the atmosphere utilizing a supersonic
ram-air ballute (combination balloon/
parachute), after which a steerable
parachute, guided by GPS, returns it to
the landing site. For manned missions,
they anticipated doing a final engine
firing to take out any translation (back
and forth or sideways) velocities and to
soften the landing. Unlike the horizontal
providers, their system was unpiloted
and automated. Their argument is that
with current technology, human failure
is more likely than avionics failure.[9]

Artist concept of an Armadillo
rocket with four independently
throttled engines

So what could go wrong? Well, as
with the others, an engine could blow
up. Armadillo has had a pretty good record in this regard, but
again, it cannot be ruled out. If the engine shuts down prematurely,
the parachute could deploy and return it safely. Presumably the
(liquid) propellants could be dumped, but if not, and the parachute
isn't sized for landing with a full load, then a hard landing would
result, with perhaps catastrophic consequences for the vehicle's
inhabitants.

Once in space, if the ballute failed to deploy, or came off, the
entry would be too fast, and probably uncontrolled, and likely
result in a total loss of vehicle and contents. Same applies to the
main steerable parachute—failure to deploy, or to keep it attached,
would result in a very hard and probably catastrophic landing.[10]

---

9 As Armadillo Aerospace VP & Project Manager Neil Milburn noted in an email to
the author, "Computers don't go out on a bender. the night before or have problems
with their [family]."

10 In late January of 2013, on a test flight, the ballute in fact detached from the
vehicle, and the resulting high entry speed tore off the landing parachute as well. This
resulted in a high-energy impact with the ground and loss of the vehicle.

A loss of cabin pressure could result in injury or death, unless the passenger is suited. As far as I know, they hadn't decided whether or not to follow XCOR's lead in that regard. But at least they wouldn't have to worry about incapacitating a pilot in such an event. However, they would have to worry about a failure of the avionics that controls the system, if it's not redundant.

What Should The FAA Do?

How likely are any of these failure scenarios? Some more than others, most of them not very. But you can't rule any of them out. In all cases (unlike NASA) there will be extensive flight tests before paying customers will be allowed aboard, because (also unlike NASA) they will be able to afford them, given the low marginal cost of a flight with a reusable vehicle.

The Federal Aviation Administration (FAA) Office of Commercial Space Transportation (FAA office code "AST") is responsible for licensing commercial launches. AST is the FAA's only space-related line of business and was established to:

- Regulate the U.S. commercial space-transportation industry, to ensure compliance with international obligations of the United States, and to protect the public health and safety, safety of property, and national security and foreign policy interests of the United States;

- Encourage, facilitate, and promote commercial space launches and reentries by the private sector;

- Recommend appropriate changes in Federal statutes, treaties, regulations, policies, plans, and procedures; and

- Facilitate the strengthening and expansion of the United States space-transportation infrastructure.

Given just the three approaches for spaceflight presented above, what rules would FAA-AST impose for the purpose of "safety"? That vehicles must be piloted? That they must be redundantly piloted? That they must be automated? That they must have automated backup? That the pilots must wear suits? That rocket motors must be liquid? Or solid? Or hybrid?

Unlike the rest of the FAA, FAA-AST still has a responsibility to promote the industry as well as to ensure public safety (emphasis on the "public"). As long as the moratorium remains in place, it has no statutory authority to protect passenger safety, at least until there is a significant accident, at which point it would have more discretion.

Some will argue that part of promoting the industry is to ensure that it doesn't kill its customers, but the industry already has ample incentive to not do that, and the FAA isn't any smarter on that subject than the individual companies within it—everyone is still learning.

There is a popular view in the space community that the first time someone dies in private spaceflight it will somehow doom the industry. Bluntly, I believe that is nonsense, because it is based on absolutely no evidence. In fact, there is an abundance of counterevidence with examples being the early aviation industry, various extreme sports including free diving and mountaineering, and even the recent cruise-ship disaster of the *Costa Concordia*, in which at least thirty passengers died.

In fact, it may take just such a death, a sanguineous christening, to normalize this business, and end the mystical thinking about it.

It is in fact a sign of our lack of progress and ambition that more people aren't killed in the quest for space. For clarity, of course, I don't mean that we should be *trying* to kill people (which would be trivially easy in this business), but rather that we should be striving to increase the level and pace of activity such that some number of people dying in the partaking of it becomes inevitable (just as tens of thousands die in auto accidents each year).

That's the practical reason that it is premature to regulate this aspect of the industry.

But to get back to the second reason that overregulation, (or any regulation whatsoever), of passenger safety would be a mistake, philosophically, we should also recognize that different people have different risk tolerances for different payoffs. Moreover, some see risk as a reward in itself (e.g., extreme sports enthusiasts, who get an adrenalin high from it). For example,

The wingsuit

there is a form of extreme BASE jumping (parachuting) with "wingsuits" (there's a popular venue for this in Norway) in which people jump off cliffs, reach speeds of up to 150 mph, and pop the chute just in time. It actually requires a drogue to pull the main chute. In March of 2013, two wingsuiters did this in Rio de Janeiro, Brazil, dropping from ultralight aircraft, with one of them swooping between two adjacent skyscrapers at such a speed. There have been about fifty fatalities from this sport over the past decade.[11]

"Free diving," in which divers see how deep they can go without equipment (often hundreds of feet) kills dozens a year, usually from drowning after a "shallow-water blackout" on the way back up (about a 2% rate). Nowhere (or at least in very few countries, but I am aware of none, other than those like North Korea, in which almost everything that is not mandatory is forbidden) is this against the law.[12]

Mount Everest is literally littered with the corpses of those who fatally failed in their attempts to conquer it. One of the most striking memories that climbers of the mountain gain from the experience (if they survive) is walking past the perfectly preserved bodies on the way to and from the summit. In some years about 10% of those climbing it don't return, though the average rate has been reduced to about 2% in more recent times, not counting support teams. In that regard it's worth noting that NASA Shuttle astronaut Scott Parazynski made two attempts, the second one successful, but he didn't do it until the year after he retired from

11 "Fatality Statistics," *BLiNC Magazine* (BASE Jumping), http://www.blincmagazine.com/forum/wiki/Fatality_Statistics

12 "Free Diver Dies Trying to Break World Record" *ABC News*, October 17, 2003, http://abcnews.go.com/GMA/story?id=125625&page=1

NASA, and both attempts were much more hazardous than any of his five Shuttle flights, even including his four dangerous space walks (extravehicular activities, or EVAs) on STS-120. As space entrepreneur James C. Bennett noted in comments at my blog in early 2012, "It is still the case that the least safe way of getting to 29,000 feet is to walk there. Yet it is nevertheless legal to do so."[13]

But perhaps the best example is the event of October 14th of 2012. In eastern New Mexico, Austrian skydiver Felix Baumgartner put on a pressure suit and climbed into the gondola of a balloon of more modern materials and much larger than the one used by Joe Kittinger over half a century before, and ascended to a record altitude of 24 miles (128,000 feet), to beat Kittinger's decades-old mark. (Interestingly, Kittinger himself acted as "Capcom" or capsule communicator for Baumgartner's mission). As he passed up through the troposphere, the face mask of his helmet started to fog up from the cold, due to a lack of sufficient power to heat it. Had it continued, he would have had to jump blind, and rely on his radio for someone to tell him when to open his chute, because he wouldn't have been able to see his altimeter. But they switched to another power source, and the frost cleared. When he jumped, after a couple minutes of free fall he went supersonic and beat Kittinger's speed record, but then went into an uncontrolled tumble for several seconds that he later related "terrified" him. He stabilized himself, deployed his drogue chute, and came in for a landing with his modern parafoil that allowed him to take four short steps into the record books. The FAA was obviously involved with, and concerned about this flight. They had to provide clear air space for the balloon to ascend and for the jumper and gondola to fall, and issue a notice to airmen (NOTAM) to ensure that no uninvolved person on the ground or in the air was injured or killed, or property damaged.

But just as FAA-AST in its licensing process has no involvement with mission success for a satellite launch, the aviation branches of the FAA had no involvement with the safety of Baumgartner, because that was not their responsibility (skydiving is a special category—even skydiving from the edge of space).

---

13 http://www.transterrestrial.com/?p=41195

Baumgartner's jump was just the beginning of a possible new extreme sport, which will not just provide thrills and entertainment (as the barnstormers once did), but also push the technology of pressure suits and wing suits, ultimately perhaps with jumps all the way from orbit. This will contribute to the future safety of space adventurers by teaching us how to survive space disasters, and provide better equipment for them. The next step will be space diving, with dives all the way from a hundred kilometers altitude, commonly recognized as the beginning of space.

Armadillo has stated that this is one of the potential markets for their vertical ascent/descent vehicles, which they've characterized as an "elevator to space" (though not orbit, at least not yet). If the aviation FAA (which is focused on safety these days) doesn't involve itself with the well being of someone jumping from a balloon, why should FAA-AST involve itself with someone jumping from a rocket? And if they don't, why should they involve themselves with the safety of *anyone* riding a rocket, at this point?

Or consider the case of the *Inspiration Mars* plan. In order to get the crew and hardware needed to perform the mission into orbit, they will need to get a launch license from FAA-AST (and probably more than one, since it seems unlikely that the mission will be feasible with a single launch, even with Falcon Heavy). That license will be issued after reviewing it to ensure that it is not deleterious to the security or national interest of the U.S., or a danger to (among other things) the environment or public safety. To what degree should they have a legitimate concern with the safety of the mission participants themselves, given that this will be a non-government mission?

On ascent, none at all. Entry, however, could be more problematic. This vehicle will be the largest, fastest artificial object to ever return to earth, at over fourteen kilometers per second (escape velocity is a little over eleven). Energy goes up directly with mass, and as the square of the velocity, so it will have about sixty percent more energy than an equivalent mass coming back from the moon (e.g., an Apollo capsule). If, instead of skimming the top of the atmosphere to gradually slow[14], it comes in at a very high angle, or even vertically—that is, it targets the center of the earth rather than the limb (i.e., edge), it could do as much or more damage than a meteoroid[15].

Presumably, it won't be just the capsule that enters, but the expandable module and trans-Mars Insertion stage as well, though they'll have to separate from them first in order to allow a safe capsule entry (the capsule will presumably only be designed to enter itself, not the whole deep-space assembly). If it does enter (that is, they only do a small separation maneuver to build up a little distance between them beforehand) they would burn up in the upper atmosphere, though it's possible that some parts of them might make it to the ground. The other option would be to separate while on a trajectory that misses the earth, and then do a correction burn of the capsule to have it enter properly, leaving the rest to become one or more new artificial near-earth objects orbiting the sun. This would involve a trade between the risks of pieces of them hitting the ground and those of creating new extraterrestrial navigation hazards, and the risk of killing the crew if they fail to perform the final entry maneuver, sending them off for another solo journey around the sun without sufficient supplies. From a crew-safety standpoint, aiming for a good entry interface from millions of miles out, minimizing the time that they have to spend in the more-cramped capsule, would obviously be preferable, but it increases the risk to everyone else, even if by a minuscule amount.

As noted in the previous chapter, the safety of the crew for this project is a primary mission goal. If they don't survive, the mission will have failed, with little to show for it except perhaps lessons learned for future attempts. The crew is an essential element for

---

14 The current mission baseline is to make two passes, reducing it to seven kilometers per second for each pass with a several-day earth orbit in between them. This would reduce the heating energy per pass by seventy-five percent, though it increases the risk of bouncing off the atmosphere into deep space, which would be fatal for the crew.

15 The one that exploded high over Chelyabinsk in Russia in early 2013 would have destroyed the city had it released the energy at a lower altitude, though due to its much smaller mass and velocity, the energy of the capsule would be more than a thousand times less.

hardware success, in that they must maintain the life-support and other equipment necessary to get to Mars and back. The question is: will they also be necessary to ensure a safe entry? That is, will they be responsible for the adjustment burns needed to get the correct entry interface? Or will it be done from the ground?

If they are responsible for maneuvering, then the FAA will have a legitimate interest in ensuring that they are alive and healthy at least through the final correction burn to ensure an entry that poses no risk to those on the ground. If not, then the agency has no current statutory authority to be concerned with whether or not they return dead or alive, any more than they are concerned with whether or not a commercial payload gets to orbit. That is, (and to repeat) their sole job with regard to safety is *public* safety, not mission assurance.

Now, we have a wealth of decades of demonstrated experience since the early sixties with performing very precise and accurate planetary maneuvers with unmanned probes[16], so in fact there would be no reason to put the crew in the control loop, and this would be a good reason not to do so, if the project's planners want to minimize potential issues with their launch licensing in that regard. If the crew is not mission critical for entry then (at least prior to the current planned end of the moratorium in October, 2015, and if the prospects for extending the moratorium look poor, it would behoove them to get a launch license before that date) the FAA would have no legitimate concern with their safety. It would be well not to establish a precedent in which they did.

In an informed-consent legal regime (which is the current one), what is important is not the level of safety *per se*, but how well informed a participant is of just how safe (or unsafe) it is. It is then the individual's responsibility to accept that risk or not to accomplish their goals. Like regulations demanding vehicles of a specific design type, a "one-size-fits-all" safety regime would not just stifle innovation, but the freedom of people to hazard their

---

16 A notable exception was the loss of the Mars Climate Orbiter in 1999, when it did a "controlled flight into terrain," striking the planet instead of orbiting it, due to a confusion between English and metric units. Its trajectory into the planet was, however, quite precise. It should be a valuable lesson for Inspiration Mars.

own lives in the pursuit of their dreams. That was a founding principle of America, another frontier. In that sense, to impose such a regulatory regime on the industry today would be in fact un-American.

What Should Industry Do?

There may be some things that the industry itself might want to do on its own, modeling analogs from other industries. The Commercial Spaceflight Federation (CSF), a commercial-spaceflight industry-advocacy organization, and the American Institute of Aeronautics and Astronautics (AIAA), the technical society for the aerospace industry in general, may want to get involved. For reasons previously stated, the sort of prescriptive regime that is the current one for aviation should be put off as long as possible because this is a regime that only works for a very mature industry, and can actually result in hazards when imposed inappropriately. For instance, the Titanic had only half the number of lifeboats needed because the prescriptive rule at the time was that it needed only sixteen (it had twenty), based on gross register tonnage, rather than passenger capacity, a rule that hadn't been updated for almost twenty years, failing to anticipate such future large passenger ships. A more goal-oriented, performance-based approach would be to not prescribe the number of lifeboats on a rote formula, but rather demand that there be sufficient means to safely evacuate all of the passengers in the event of an emergency, and leave it up to the operators to determine how best to do this.

Space lawyer Michael Listner, president and interim CEO of the International Space Safety Foundation, and others have recently suggested that in fact the maritime industry, more than aviation, provides a useful model for such self-regulation, in privately developing recognized standards that can act as a defense against litigation in the face of adverse events.

Certification against goal-oriented and performance-based standards is complex, and requires a more complex argument and evaluation strategy. To verify compliance of goal-oriented

(Quote continues)

and performance-based standards, the safety certification team must have a deep knowledge of how the system is meant to work in order to understand the relevant hazards and the soundness of the design controls selected to mitigate the risks. In principle, the safety certification team should be even more knowledgeable and experienced than the design team. This is currently the case with NASA's safety certification teams, who evaluate the design of privately developed commercial human spaceflight vehicles such as the Dragon space capsule by SpaceX. However, there are industries where relying on past experience to build a safe system is simply not possible because the system is completely new, highly safety-critical, and extremely expensive. This is the case with the commercial spaceflight industry.

Considering this, the future of regulation in commercial human spaceflight activities lies in self-regulation. The traditional government role in establishing safety regulations and certifying compliance is no longer suitable for highly advanced and fast evolving systems and operations such as suborbital and future commercial space transportation. Rather, the commercial space community should take the lead in developing goal-oriented and performance-based safety standards and policies. These companies have the necessary in-house skills and technical resources to thoroughly and efficiently verify compliance with regulations and standards. This would be a more efficient and cost-effective approach, as long as such verifications are performed with the necessary independence and rigor.

The government's role should be to define certification process rules and maintain a general supervisory and auditing role of the safety assurance and management system. This mirrors the key recommendations of the presidential commission that investigated the Deepwater Horizon oilrig [sic] disaster in the Gulf of Mexico in April 2010. In their report, the commission recommended that "the gas and oil industry must move towards developing a notion of safety as a collective responsibility."

Such a recommendation is less revolutionary than may appear, however, because the maritime industry has done precisely this for more than 250 years in regards to the design, construction, and maintenance of seagoing vessels through classification societies.[17]

Classification societies were born in the dawn of modern free-market capitalism, in the coffee houses of mid-eighteenth-century London, as ship captains and owners met to discuss how to allocate risks and rewards of shipping ventures at places like Edward Lloyd's, where the first registration society was born (as was the modern insurance industry). The societies were non-profit and non-governmental, and would accept no liability, but provided a means of establishing standards of seaworthiness that everyone could recognize and on which they would agree.

The authors propose that the commercial spaceflight industry adopt a similar approach, in which industry standards of "spaceworthiness" could be established. However, there are some key differences between the maritime and launch industries.

The first is that the latter is driven by regulations emanating from the 1967 Outer Space Treaty (OST) and the associated 1972 Liability Convention, which make all States Parties to them responsible for space-related activities originating from entities (individuals and corporations) within their jurisdiction. Ships, on the other hand, because they don't operate under such strictures, can fly registered flags of convenience, with no responsibility of any other nation, regardless of nation of ownership or design.

The second difference, which can only be rectified by time and experience, is the vast difference in technological maturity between the two industries. We now have hundreds of years of experience upon which to draw for defining shipbuilding and operating standards, whereas the first suborbital commercial

---

17 Listner, Michael J., Sgobba, Tommaso and Kunstadter, Christopher, "Taking a page from maritime practice to self-regulate the commercial space industry," *The Space Review*, March 4th, 2013, http://thespacereview.com/article/2252/1

spaceflight, and commercial orbital human spaceflight, have yet to occur. Related to this, as Listner *et al* point out, is that, given that state of technological immaturity, there is a lot of competition within the American industry that enhances their reluctance to share data even among themselves, let alone others (as previously noted, Blue Origin is the most notorious in this regard). They suggest that an independent safety institute, as exists with *Formula 1* car racing, could alleviate this to some degree, just as they are willing to reveal information to government agencies evaluating their technologies in proposals, with legal safeguards against the revelation of proprietary information.

The third difference between the two industries, at least with regard to U.S. launch companies, is the International Traffic in Arms Regulations, or ITAR, a U.S. law that restricts access to technical details of launch technology by non-American entities, making it more difficult to share knowledge in an international context even if the participants wanted to.

Maritime classification societies are subject to lawsuits by ship owners and operators if a flawed inspection results in a loss due to unseaworthiness. But the authors point out that this undermines both the purpose of the societies, and the ultimate responsibility of those same owners and operators:

> The theory of liability to ship owners rests on the presumption that the Classification Society's inspection of a vessel was not in accordance with their own rules and standards. Generally, suits by owners or insurers for the loss of a vessel allege that the Classification Society performed its inspections in a negligent manner and breached an implied warranty of workmanlike service. However, the courts have not generally looked favorably upon these actions, recognizing that a cause of action on the first basis would undermine the traditional doctrine that imposes the non-delegable duty on the ship owner to maintain its vessel in a seaworthy condition. The courts also generally recognize that allowing a cause of action in this instance could be a precedent for making the Classification Society the ultimate insurer of

any vessel it surveyed. This could have the undesirable effect of making the Classification Society ultimately responsible for a vessel's seaworthiness, which is in contradiction to the duty imposed on the ship owner.[18]

They also note that in theory, claims could be brought against a society by third parties, but there have been no such successful suits to date in the maritime industry, which should offer some comfort to those contemplating establishing one for commercial spaceflight. Similar arguments against liability would have a good chance of prevailing, with some exceptions:

> [T]he maritime principle of the implied duty of seaworthiness would not apply to commercial spacecraft frames. Furthermore, unlike the loss of a ship, the wreckage of a commercial spacecraft involved in an accident may at least partially recoverable, which might provide evidence of the cause of the accident. However, the burden would still be on the owner and the insurer that negligence on behalf of the Classification Society was responsible for the loss.[19]

There is one other thorn in the bouquet:

> [S]ince certification activities of commercial spacecraft will operate under state law as well as federal law, there is the potential that contractual concepts of warranty under state laws could apply to surveys of commercial spacecraft. Specifically, certification of a commercial spacecraft could create an express warranty under the contract law of the states where the spacecraft is built and operated. Furthermore, since Classification

---

18 Listner, Michael J., Sgobba, Tommaso and Kunstadter, Christopher, "Taking a page from maritime practice to self-regulate the commercial space industry," *The Space Review*, March 4th, 2013, http://thespacereview.com/article/2252/1

19 Ibid.

Societies would not be protected under immunity laws given to commercial operators, any accident could produce a suit from the spacecraft owner or operator based on breach of that express warranty. Classification Societies in the maritime realm do not encounter such situations.[20]

They go on to suggest that Congress could amend the Commercial Space Launch Act to extend its recognition of maritime classification societies to ones for commercial spaceflight, and grant immunity from state laws, which would at least reduce if not eliminate the potential flow of lawsuits, with a corresponding reduction in insurance rates for the industry.

Along these general lines, pioneering space entrepreneur James C. Bennett has made a more specific proposal for what he calls a Spaceflight Safety Best Practices Institute (SSBPI). Recently developed in the course of discussions with the United Kingdom Space Agency (UKSA) about how to best regulate suborbital flight in the UK, the SSBPI could generate reference standards via its stakeholders, which could then potentially be adopted after appropriate review as "safe harbor" recommendations from various regulating bodies, such as UKSA or the FAA.

The institute would deal with safety issues related to space-launch systems (suborbital and orbital), spaceports, in-space facilities, and other space-related activities and facilities, with the criteria for inclusion driven by the potential to generate liabilities associated with the aforementioned OST/Liability Convention.

He proposes that it have the following characteristics:

- Non-profit, non-governmental, funded by industry stake-holders with perhaps some minimal government funding

- Small, with few employees, consisting of administration, engineering and policy staff

---

20 Listner, Michael J., Sgobba, Tommaso and Kunstadter, Christopher, "Taking a page from maritime practice to self-regulate the commercial space industry," *The Space Review*, March 4th, 2013, http://thespacereview.com/article/2252/1

- SSBPI staff would not themselves generate the standards, but rather referee standards committees for the various segments of concern, composed of industry representatives, analogous to the Commercial Space Transportation Advisory Committee (COMSTAC) long supported by FAA-AST

- International in nature, but not intergovernmental, with offices near regulating bodies (initially London and Washington, but expanding as more nations adopt the concept)

There would be a broad array of stakeholders—vehicle builders, operators, customers, insurers, spaceport operators—to support the institute. Government agencies would not be members, or have oversight, but would participate and provide insight. He suggests that institute staff be prohibited from having any financial interest in any of the stakeholders to eliminate the potential for conflict of interest, including a requirement for a gap between leaving a stakeholder's employ and becoming staff, or vice versa. This would allow a standing non-disclosure agreement on the part of the institute, similar to that for government employees overseeing industry contractors. Additionally, while staff would mainly be drawn from the country in which the institute office was located, provisions would have to be made for "no-foreign" channels for the handling of export-controlled data and materials.

Ultimately, it could establish small research and development laboratories for developing and testing safety solutions to be made available to the industry as a whole, perhaps building on the current FAA-AST "Centers of Excellence" except outside of the government. These could eventually become additional sources of revenue for the institute.

One of the first standards that the industry might agree on could be a three-part classification of risk in general, with appropriate levels of assumed risk. The lowest is regular commercial service, which is not contemplated at present, although before too long, in the next generation or two, the same sort of vehicles used for suborbital flights may be used for suborbital transport hops (e.g., LA-Tokyo). The second level would be the sort of assumed risk employees accept in hazardous jobs—helicopter transport to North Sea oil rigs, etc. The third level would be comparable to adventure activities—such as bungee, etc., or extreme environment research, like Antarctic aviation.

All of today's space-related activities would be in the third, highest assumed-risk category for now. Within that category, it might also be useful to classify degrees of medical risk (i.e., from stress characteristics of the nominal flight environment, such as acceleration, etc., rather than likelihood of accident) assumed by flight participants. The lowest category would be comparable to common-carrier transport; the next highest would be comparable to extreme work environments; and the third, highest level, would be for assumed-risk activities like technical mountaineering or diving, which demand a very high degree of fitness and even then regularly experience losses of qualified personnel.

Table 2: Possibilities Range From Low To Very High Risk

Vehicle Risk

		Regular Commercial Service	Professional Assumed Risk	High-Risk Adventure
	Extreme Adventure		Private Rocket Packs	High-Gee Vehicles With Unpredictable Abort Locations
Health Risk	Extreme Work Environments		Suborbital Research	Vehicle Without Abort System Vehicle Without Environmental Control
	Common-Carrier Transport	Routine Point-to-Point, and Orbital Transport (Currently Non-Existent)	Virgin Galactic, XCOR Lynx, Armadillo, Blue Origin	Single-String Environmental Control

These possibilities are displayed as a matrix in Table 2, with the horizontal axis (x) being risk of loss of vehicle and the vertical (y) being risk of loss of participant. Of course, in reality not every square in the matrix would be populated. X-axis risk would be equal to all participants in a flight of a given type, while Y-axis risk could be controlled by medical exams, which would probably be required by insurers in any event. The regulatory task would be to insure that levels of risk and flight-participant medical criteria were properly characterized.

With regard to medical requirements, this will be very much a function of vehicle design and flight profile. Obviously a ride that entails high levels of acceleration will be harder on an unhealthy individual than a gentler trip. It will also depend on the level and type of environmental control within the crew cabin, and whether or not pressure suits will be required. As described above, we are already seeing a fairly wide variation in the developing industry in such factors. However, given the cost of such trips, it wouldn't be unreasonable to weed out potential problem customers by requiring training in simulators, such as the National Aerospace Training and Research Center (NASTAR) centrifuge and hypobaric chamber. Such training will again be vehicle/trajectory specific, and would be best specified by individual providers.

Similarly, the industry may also want to develop a classification system similar to the six-class system currently used for river rafting, to indicate the level of hazard. For example:

Level 1: No more than three gravities acceleration, full cabin life support within minimal temperature range and CO_2 levels for duration of trip

Level 2: No more than five gravities acceleration, full cabin life support with wider variations on CO_2 and temperature

Level 3: Potential for very high short-duration acceleration (greater than ten gravities), in a space suit for duration with minimal thermal control

Level 4: Level 3, plus high potential for abort in unpredictable location (wilderness or cold ocean)

To the highest degree possible, vehicles should be able to be autonomous from ground control, with on-board crew having full vehicle control, to avoid the risk of communications loss and lag.

However, as we've seen, some providers (e.g., SpaceX Dragon, Armadillo) are designing automated vehicles. For passengers, there will be psychological factors associated with the presence or absence of a flight crew, and many will want to have someone in control with literal "skin in the game." This is a consideration best left to the market for now.

As with other factors, the amount of flight tests will vary with vehicle design. Fortunately, as already noted, at least for the suborbital vehicles, extensive flight testing (for the first time in the history of the space industry) will be affordable. As previously discussed, there is no need for prescriptive regulations at this time, both because the industry is too immature to know what they should be, and because all suborbital companies are planning extensive flight-test programs. For orbital systems, because of the current high cost of test flights on expendable vehicles, the space agency will have to rely much more on analysis and simulation, as it has been for the COTS and Commercial Crew programs. In all cases, though, if someone is offering rides on a vehicle with insufficient testing (in the opinion of some), I would still suggest a *caveat emptor* ("let the buyer beware") approach (and a continuation of the current requirements of informed-consent), not a regulatory prescription.

With respect to orbital spaceflight, I've recently speculated on what SpaceX would do if someone were to come to it now, having successfully flown a Dragon to the ISS multiple times, even with the engine failure, and said, "Here's a couple hundred million dollars. I and my friends would like to take a ride into space, to the ISS. We understand that you haven't yet developed a launch-abort system for it, but we are confident in the Falcon as is, and are willing to take the risk." What should, what would Elon Musk's response be? From a business standpoint, it would be taking a large risk, because if he killed his customers it would make it harder to get new ones. But there are some reasons why he might want to do it:

1) Why turn down launch business?

2) By actually returning humans to earth from the ISS in the near term, the trip would truly demonstrate the ability of the vehicle to act as a lifeboat (assuming the installation of a true docking adaptor), allowing NASA to purchase it for that purpose long before it is capable of crew change

out, reducing the nation's reliance on the Russians for that purpose, and (as discussed earlier) allowing larger crew sizes on the ISS.

3) It would demonstrate the safety of the system for carrying crew, even without a launch-abort system. The LAS will be a safety enhancement, but it is not critical for flying people. It will also shut up the critics who complain that SpaceX has never flown anyone and hence can't be trusted to fly NASA personnel.[21]

4) It would allow them to maintain control of their own safety criteria for non-NASA customers, and prevent NASA from establishing a very costly *de facto* industry standard with their requirements for their own personnel. It would also make the crucial point (as previously stated) that different people have different risk tolerances, depending on the reward, and that the government should not determine the level of risk that private individuals are willing to accept.

In response to the recent Inspiration Mars announcement, the company stated that although it does not have an "official relationship" with the Inspiration Mars Foundation, it "will always consider providing a full spectrum of launch services to interested customers."[22] So frankly, based on his own philosophy and things like this and others that he's said in the past, I'd have to think that Elon Musk would be very tempted, despite the storm of criticism that would likely greet the decision. No doubt some people would call him "reckless," and unfit to have the precious lives of our astronauts entrusted to his company. Here's a hypothetical speech that I can imagine he might give to explain it:

Some say that SpaceX is being "irresponsible" by providing a space flight to willing adventurers on a vehicle without a back-up abort

---

21 There are always some people who declare that nothing can ever be done for the first time.

22 Morring, Frank, "Serious Intent About 2018 Human Mars Mission," *Aviation Week & Space Technology*, March 4th, 2013, http://www.aviationweek.com/Article.aspx?id=/article-xml/AW_03_04_2013_p24-553876.xml

system. And it would indeed be most irresponsible for us to fly these daring people on the pretense that their ride will be risk free.

But in fact, we would never so characterize a ride on one of our vehicles, even if the launch-abort system were complete and fully tested. It would be equally irresponsible for an airline to claim that its flights were risk free, or that it was risk free to drive a car to the grocery.

Or to get up in the morning.

It is easy to say that every trip should be made "as safe as possible." But if we really meant that, we would never travel at all. After all, if you drive at seventy miles an hour, you aren't traveling "as safely as possible," because you could be going sixty instead. Or fifty. But even that speed can kill if you run into something at it, so perhaps we will have to back down to five miles per hour, the speed that bumpers are designed to, if we want to be really safe. But even then, there is still the risk of some other car traveling at seventy hitting you broadside or head on. So the safest thing to do is to not go at all. That would be "as safe as possible," though not safe as practicable.

Early in the last century, John Shedd wrote the words, "a ship in a harbor is safe—but that's not what ships are built for." Neither are our space ships built for languishing in hangars or on launch pads because some fear to use them, particularly when those who fear are not those who risk.

We have confidence in this system, as do many satellite customers, because they have entrusted us to carry their satellites worth hundreds of millions of dollars on it, and there is no back-up system for them, but we recognize that something can always go wrong, just as it could go wrong with the abort system as well.

While NASA is now claiming to demand the highest possible levels of safety for their astronauts, despite the fact that for three decades they themselves flew a vehicle without an effective launch escape system, that doesn't mean that levels below that are "unsafe" in any binary sense. Every decision we make, every act we take, from getting out of bed in the morning, going in to work, to taking our next breath of air, involves some level of risk. There is no absolute safety, this side of the dirt. But we engage in all those acts, with whatever level of risk they bear, because we anticipate some reward for them. Not everyone is NASA, and while we are willing to meet NASA's requirements for NASA's purposes, for a price, it is not in our or the industry's interest to allow that to become a de facto standard of safety—establishing one for this industry is, with all respect, not NASA's job.

I ultimately came to this country from South Africa because it represented freedom, and that included freedom to take risks, and freedom to fail. I took advantage of that freedom to create a new company that would allow others to do the same, on this new high frontier. Everyone has their own tolerance for risk, and willingness to pay for it, and everyone in a free country should be able to engage in activities, however risky, that potentially harm only themselves.

If some believe that the potentially higher level of risk of death in a flight without a launch-abort system is worth the reward of the adventure and the history, it would be hypocritical of me and my company to stand in their way, given the risks we have taken ourselves in achieving our own success to date. Enabling people to go into space, to pursue their own dreams, at their own informed risk, is exactly the reason that I founded this company.[23]

---

23 I have never discussed this topic with Mr. Musk. These words are mine and mine alone.

Chapter 10

Conclusions And Recommendations

We do not know where this journey will end. Yet we know this: Human beings are headed into the cosmos.
— President George W. Bush

At the conclusion of the Prologue, I wrote that NASA's and Congress's risk aversion paradoxically increases risk and even pointlessly kills people. They do so by remaining stuck in an archaic paradigm, appropriate for its time half a century ago, but now tragically outdated. That is, the government owns and operates hyper-expensive, partly or fully expendable, experimental spaceflight systems as though they are operational, while not being able to fly them enough to make them reliable or affordable.

In their futile attempts to avoid risk, they cast into concrete decades-old flawed and hazardous ways of doing things, and also avoid the innovation and new approaches necessary to ultimately develop safer and lower-cost systems. In doing so, they also hold us back from much grander accomplishments in space, and particularly from opening it up as a new frontier and economic sphere for humanity.

Part of the reason for this, of course, is that the nation has not decided, formally or culturally, that space is a frontier for settlement and economic development or, if so, that it is NASA's role to help open it. In fact, we have never really even had a national discussion on the former subject. Yet, the notion that NASA should not only be involved, but that it should be the lead on any space-related effort, is unjustifiably taken for granted.

When I heard president George W. Bush's speech in 2004, announcing the Vision for Space Exploration (VSE, described in Appendix A), this was the phrase that stuck out to me in the context of history: "We do not know where this journey will end. Yet we know this: Human beings are headed into the cosmos."

This was a message that was implied by Kennedy's Apollo speeches, which resulted in a generation (mine) realizing that it was a false promise when the Apollo program ended. Although Apollo achieved its objective of winning a crucial (or at least, so it seemed at the time) Cold War battle against the Soviet Union, it was not really about space exploration at all. But the promise of us going out into the cosmos was never explicit, and after Apollo, NASA was reined in to low earth orbit, at least as far as humans were concerned, by the Shuttle and ISS. The first president Bush tried to change this in 1989 with the announcement of the Space Exploration Initiative (SEI), but it was still-born, never having won the support of Congress (or for that matter, NASA itself). In fact, throughout the nineties, Congress expressly forbade NASA from any serious planning for human missions beyond earth orbit.

So in 2004, when George W. Bush spoke those words and Congress accepted them as policy, it was a sea change for U.S. space policy. To me, they are at the core of the VSE, and the rest of it—2010 Shuttle retirement, Crew Exploration Vehicle, moon by 2020—are details.

However, by deciding to waste time and money on new rockets that weren't needed, NASA under Mike Griffin got the details wrong, and we lost half a decade. But with the new space policy announced by President Obama in early 2010, the fundamental goal—that we humans are heading into the cosmos—remains, despite all of the hysterical cries that he was "destroying human spaceflight." At the time, Laurie Leshin, the new head of the NASA Exploration Directorate, agreed and in 2010, she gave a speech[1] in which she declared that "the goal remains the same." In my opinion, the Obama administration's new approach, while far from perfect, has a much better chance of making it happen and is much more in keeping with the original criteria of the VSE that were set out by the Aldridge Commission (of which Dr. Leshin was

1 Leshin, Laurie, keynote speech at the Marshall Institute, March 30, 2010, http://www.c-spanvideo.org/program/292791-1

CONCLUSIONS AND RECOMMENDATIONS

a member), that was put together after the announcement of the VSE to provide guidance on how to implement it: that activities associated with it be affordable and sustainable, and support commercial and international participation. These are criteria that Constellation, Orion and SLS clearly did not and do not meet.

If settlement and development are not the purpose, then other than for the dubious (given the cost) purpose of national prestige, it's unclear why we have a government human spaceflight program at all. And when purposes are unclear, the justification to risk, let alone sacrifice human life for it is equally unclear, and hence the national aversion to it.

One of the most significant results of the Augustine Committee in 2009, made all the more so by the lack of reportage on it in the media, was its very clear statement that "the ultimate goal of human exploration is to chart a path for human expansion into the solar system."[2] Jeff Greason, a member of the body that made that declaration, has said that settlement is an implicit national goal:

> "We have just started, I think, to realize in the last eight or ten years that we do have a goal for the national space program," he said. "There is a national consensus among policymakers that we have that goal, but everybody's kind of afraid to say it, because they're not sure we can do it." That goal, he said, is the "s-word": "It is actually the national policy of the United States that we should settle space."

> To support that claim, Greason cited a number of studies and speeches, including the conclusion of the Augustine Committee [about the ultimate goal of human exploration]. He also noted President Obama's speech at the Kennedy Space Center last April included a veiled reference to space settlement: "Our goal is the capacity for people to work and learn and operate and live safely

---

2 Augustine, Norman, et al., "Summary Report of the Review of Human Space Flight Plans Committee," October 22nd, 2009, http://www.nasa.gov/pdf/384767main_SUMMARY%20REPORT%20-%20FINAL.pdf

beyond the Earth for extended periods of time, ultimately in ways that are more sustainable and even indefinite."[3]

He could have mentioned also the Space Settlement Act, passed by Congress in the late eighties, that legally required NASA to regularly report on its progress toward that goal. But it was a toothless law that the agency steadfastly ignored for years.

The policy road ahead remains uncertain, as always, but with acceptance by first a Republican, and now a Democratic administration and Congress, the big issue of space policy, that it is the goal of this nation to settle space, seems politically resolved. In that respect, we are in a better position than ever in our nation's history to finally get serious about that goal, despite Congressional attitudes of pork over progress.

But if settlement and development are the goals, one would never know it from our space policy or discussion about it in the halls of Congress, NASA or the White House. In no way is this more the case than in our dysfunctional national attitude toward space safety described in the preceding chapters of this book. Despite the myths of the Apollo program, there is a lot more to charting a path to open a frontier than to occasionally and at great expense send a few government employees out to briefly stroll on another planet. To imagine otherwise would be as though Lewis and Clark had gotten no further than St. Joseph, Missouri. Or that they were the beginning and the end of the opening of the North American continent, (though many others both preceded and succeeded them, many of them privately, for private reasons).

While the primary purpose of the initial long screed that turned in time into this book was to get a national and political conversation going on our approach to safety in human spaceflight, an intrinsically necessary part of that discussion must be the purpose(s) of same. We cannot determine the appropriate level of requirements for our space systems, either government or private, if we don't understand why we're building and operating them.

---

3 Foust, Jeff, "New strategies for exploration and settlement," *The Space Review*, June 6th, 2011, http://www.thespacereview.com/article/1860/1

If this book succeeds in its goal of kicking off that broader discourse, and we finally make explicit the implicit policy of settlement described above, then we will need to decide how best to restructure federal policy to make more rapid progress toward that goal. The aim of this final chapter is to provide some suggestions to that end, some of which will be rightly viewed as a radical departure from the policies of the past decades, but that makes them neither wrong nor unnecessary. They need not involve the expenditure of any more taxpayer money than we are currently spending, and they could perhaps be done for even less. But they do require that the money be spent much smarter. I hope that these suggestions will provide further fodder for the intelligent national discussion on human spaceflight that has sadly eluded our nation since the time of Apollo.

Commercial Spaceflight

Ultimately, the keys to opening up space to humanity are the development and utilization of extraterrestrial resources, and affordable access to orbit for humans. If we have learned anything from the past half century of human spaceflight, it should be that the latter will not occur until we have multiple competing commercial space transportation providers to continually improve services and drive down costs and prices, just as we have in every other mode of transportation in the free market. But in addition to competing on cost, they will be doing so on passenger safety and trip reliability as well, and the range of acceptable risk must be allowed to vary for different applications, customers, and vehicle types.

We are in a very exciting period in the development of space transportation, with a "Cambrian explosion" array of new approaches to drive down the cost of space access, similar to the early days of aviation, even before designs had settled into what is now the familiar look of a conventional airplane. Few customers will be under any illusions that such rides are as safe as commercial aviation or even the more hazardous travel by automobile, and (as in the early days of barn storming) many will be willing to accept the risk for the reward. As previously discussed, in order to meet its requirement to promote the industry, FAA regulations will have to be flexible and accommodating of this variety, lest we strangle a fledgling industry in the nest. I would make the following recommendations to Congress:

- The moratorium on regulating passenger safety put in place in 2004, and recently extended to 2015, should be extended far beyond that, to perhaps a decade beyond the first operational flight of a commercial suborbital vehicle, or even indefinitely.

- Per the advice of Listner *et al*, the Commercial Space Launch Act should be amended to extend the traditional recognition of maritime classification societies to similar ones for spaceflight and spaceports, and made exempt from state law, as shipping is.

- Eliminate the funding for the vastly overpriced Space Launch System (SLS), whose schedule continues to slip, which numerous studies indicate is unnecessary, is too expensive for trips beyond earth orbit, and which will fly so seldom that it cannot be flown safely or reliably.

- Devote more resources to purchase crew transportation from multiple vendors in the private sector for any and all of the government's human spaceflight needs. This will not only drive down the cost for both the government and private customers, but also assure that we are never again without the capability to get Americans into space on American vehicles.

Even with the end of the moratorium, the FAA should continue to regulate passenger safety with a light hand, and it should maintain a separate vehicle class, analogous to experimental aircraft, for those who want to continue to push the performance and innovation envelopes on an informed-risk basis. Outside of this class, it should establish a value for a passenger's life, within the established guidelines of one to ten million dollars, per OMB Circular A-4[4], to allow it to make rational decisions about how

---

4 Regulatory Analysis: OMB Circular A-4, http://www.whitehouse.gov/omb/circulars_a004_a-4/

much to demand that a vehicle provider spend to prevent passenger deaths, and to allow the providers to make rational decisions as to the best way to make expenditures in the furtherance of that goal.

The administration could help as well, without an act of Congress. The Office of Commercial Space Transportation (OCST), currently located within FAA and often referred to by its internal code designation as FAA-AST, should be taken from under the FAA administrator and reconstituted as a separate office of the Department of Transportation (DOT), reporting directly to the Secretary of Transportation as do normal DOT agencies. The OCST was originally constituted as a special office within the Office of the Secretary of Transportation as an interim measure under President Reagan's executive order of 1983, codified by the Commercial Space Launch Act in 1984. It was administratively moved under the FAA early in the Clinton administration as a result of Vice President Gore's "streamlining government" initiative. Reversing this, and once again giving it independent status within the department would both elevate the national importance of space transportation, and remove it from the routine-transportation, common-carrier-oriented safety environment of the FAA, which (as previously described) lost its role in the promotion of the aviation industry in the wake of the ValuJet crash in the late nineties.

As previously discussed, industry should take the initiative and set itself on a path toward self regulation via the aforementioned classification societies, perhaps starting with Bennett's suggestion of a Spaceflight Safety Best Practices Institute, and start to establish the risk and class categorization process described in the previous chapter. Industry should also, via the Commercial Spaceflight Federation and others (such as space advocacy groups), start lobbying Congress and the administration for the above.

Government Spaceflight

The phrase "human rating" should be purged from both the technical and policy vocabulary, because it means too many different things to too many people, and is fraught with historical baggage going back half a century. Personnel working in Safety and

Mission Assurance (S&MA), in both the government and industry, should similarly purge from their vocabularies the words "safe" or "unsafe" as absolute unmodified terms (such as, "we won't fly until it is safe" or "the vehicle is unsafe"). Absent quantification, both words are meaningless. Instead, we should talk only in terms of probabilities of Loss of Crew, or Loss of Passengers.

NASA must establish a finite value for an astronaut's life. In practice, the amount can't be infinite, since the agency has a finite budget and finds it necessary to occasionally get things done. If there is a political obligation to pretend that it is infinite, this means that honest discussion of safety trade-offs and priorities is effectively forbidden. This actively increases risk because, as we saw with the Ares I program, it essentially guarantees that money and effort will be allocated haphazardly, rather than being focused on the most serious safety problems. NASA is not a regulatory agency and it is not dealing with public lives, so it is not confined by the OMB circular referenced above, and could thus establish a higher value than ten million dollars. Bob Zubrin has suggested fifty million, to account for training and other factors. I personally think that's high, if we want to affordably explore space. But they have to choose something and get Congress to accept it.

The agency would do well to take a page from the United States Coast Guard, whose informal motto is "You have to go out, but you don't have to come back." But NASA's history, culture and charter probably render it intrinsically incapable of such an attitude. Therefore, it might be better, from a policy standpoint, to return NASA to the R&D function it performed as the NACA (before it was perverted into an operational agency during Apollo), and get it out of the human spaceflight business other than for technology development. Surprisingly, this would be entirely within NASA's charter, which says nothing at all about sending people into space, let alone beyond earth orbit.

To the degree that the U.S. government is involved in human spaceflight, it would be better to provide legislation that would actually establish an entirely new organization, and a complete restructuring of the federal space establishment, including military. Now that the Cold War is long over, there is much less need to have a sharp delineation between military and civilian space activities. We might want to reboot our space policy on more familiar maritime models.

Building on an earlier proposal by U.S. Air Force Lt. Col. Cynthia A. S. McKinley, space entrepreneur and consultant James C. Bennett has proposed that a new organization be stood up with a charter that will allow it to be (among other things) less risk averse. Modeled on the Coast Guard, it would be a uniformed service that reported to a civilian department (as the Coast Guard used to report to the Department of Transportation before it was absorbed into the new Department of Homeland Security), and be attached to the Air Force as the Coast Guard is to the Navy, in times of war.

A U.S. Space Guard (USSG) could perform space-related functions such as navigation support, search and rescue, space situational awareness, constabulary duties, and perhaps even regulation, among others, that need to be done, but are currently not being done well, or at all, because they are assigned to other agencies where they are peripheral to those agencies' main purposes and functions and often suffer from inattention and low priorities.[5] The USSG would have its own academy and, like the Coast Guard and unlike NASA, such a quasi-military organization would be better suited to both focus on those critical tasks, and be willing and able to risk the lives and health of the guardsmen (and women) to accomplish them, without worrying about counterproductive lectures on safety from uninformed congresspeople.

A New National Resolve

If we truly decide that we want to open up this harsh but next frontier, it is past time to take a realistic and rational approach to space safety, and space policy in general. A Congress and a president who were serious about it would forthrightly declare frontier opening a national priority, and implement the reforms

---

5 Bennett, James C., "Proposing a 'Coast Guard' for Space," *The New Atlantis*, Winter, 2011 http://www.thenewatlantis.com/publications/proposing-a-coast-guard-for-space

suggested above. A commenter at my blog, Paul F. Dietz, once suggested that a serious government would authorize and dedicate with a bold but sober and somber presidential speech a (large) national cemetery like Arlington, to provide an honored and eternal home (even if only in spirit, because many won't return to earth in any physical form) for those who will die in pioneering the high frontier. If we were really serious, every time a policy maker talked about safety being "number one," or "the highest priority," or "paramount," they would be called on it, either by hearing witnesses, or the media. If possible, they would be compelled to agree that this makes everything else, including actually doing things in space, a lower priority, with all its implications. We wouldn't continue to wait in perpetuity for the mythical, magical, giant, "safe" rocket before we started to head out to other worlds.

Once Columbus showed the way, fortune seekers and settlers didn't wait for shipboard clocks, or steam engines, or steel hulls. They set sail for the New World with what they had. A century or so ago, Rosemary and Stephen Vincent Benét wrote a poem about the days of sail, whose first stanza was:

> There was a time before our time,
> It will not come again,
> When the best ships still were wooden ships
> But the men were iron men.

As projects like Inspiration Mars and Mars One demonstrate, there are iron men and women of this age as well. All that is needed to unleash them on the solar system, so that they can blaze the trails for the rest of us, is simply for our culture and politicians to act, for the first time since Apollo, as though space is important.

Opening the Space Frontier, The Next Giant Step

This is a portion of the 71 x 16 foot mural painted by Robert McCall for NASA in 1979. It is on display at the Johnson Space Center in Houston, Texas. Flanking astronaut John Young, who flew on Gemini, Apollo and commanded the first orbital flight of the Space Shuttle, are Mercury and Gemini astronaut, Gus Grissom and Space Shuttle astronaut, Judy Resnik. Grissom and Resnik died in opening the frontier, though Young is still with us. Here we recognize that the quest to conquer the unknown is all-too-often accompanied by tragedy. Yet, as exemplified by our predecessors, we should press on.

Acknowledgments

Many people made this work possible by donating to the *Kickstarter* projects that funded it. Major donors and matching contributors were Jeff Greason, Stuart Witt, John Walker, Michael Kelly, Debora Johnson, Gregory Hill and Yves-A Grondin. I've listed in the next section those who made the necessary contribution level to be noted in this work, and I am very grateful to all other contributors as well.

I want to thank those who reviewed and provided feedback, corrections and suggested additions to early and later versions of this as it evolved through the fall and winter of 2012. Notably, they are Clark Lindsey, Jim Bennett, Robin Snelson and Jorge Frank. I want to thank Henry Spencer for keeping space history accurate, Henry Vanderbilt for his suggestion for the prologue and Bob Mecoy for his suggestion that ultimately evolved into the title of the book. I also want to thank Robin for her crucial whip hand in keeping me going on it through the summer of 2012. Most of all I want to thank Bill Simon for his tireless and invaluable efforts in designing, illustrating, editing and indexing this book. All remaining errors are, of course, my own.

Critical Sponsors and Honored Donors

This book was made possible, in part, by two Kickstarter crowdfunding projects in the spring and fall of 2012. The first funded the research and writing;[1] the second funded the design, illustration, editing, indexing and other activities up to the actual printing of the book.[2] As a promised reward for their contributions, donors to the first Kickstarter project receive electronic or autographed printed copies of the final product; donors who met the contribution-level criteria of the second Kickstarter project have their names listed here.

The list is arranged in two categories: 1) Critical Sponsors of the second Kickstarter project; and 2) Honored Donors to the second Kickstarter project. And, of course, thanks to all who contributed regardless of amount or desire for explicit rewards.

Second Kickstarter Project
To Prepare for Publication

Critical Sponsors

Dale Amon	John Walker	Donald T. Parker
Jake Casey	Chris Kaiser	Robert E. Ray
Sam Dinkin	Chris Kelsey	Glenn Reynolds
Stephen Fleming	Michael Laine	Andrew Rush
John Gebauer	Stuckey McIntosh	Eric Weder
Rich Glover	Ed Minchau	
Gregory Hill	Ryan Olcott	"Backer Name"

---

1 http://www.kickstarter.com/projects/1960236542/rationalizing-our-approach-to-safety-in-space/

2 http://www.kickstarter.com/projects/1960236542/safe-is-not-an-option-our-futile-obsession-in-spac

Honored Donors

Mark Atwood
Danny Bannister
Dmitry Bera
Matt Blackie
Joe Blowers
Paul Bohm
Walt Boyes
Byron Cain
Henry Cate
Florent Charpin
Matthew Chisholm
William Dale
Calvin Dodge
L. Doerr
Allen Edwards
Austin Epps
Carsten Erickson
Joseph Federer
Brian Lee Gnad
Susanna Griffith
Karl Hallowell
Russell Howard
Ginger Huguelet
Anne Hudson
Glen Ivey
Leland Jackson
Doug Jones
Kaido Kert
Daniel L. Kramer
Gregory Lange
Joe Latrell
Rick Lentz

Steven Madere
Josh Mandir
Charlie Martin
Thomas Matula
Robert V. McBrayer
Scott Merryman
David Miles
Jason Miller
Scott Norman
Mark Perneta
Mike Puckett
Andreas Rauer
Patrick Ritchie
Steven Rogers
Logan Russell
Ian Schmidt
Mark Shepard
Frank Smith
Kevin Smith
Ian R. Solberg
Alan Stern
Roy Stogner
Jack Stokes
Mitchell Surface
Tomas Svitek
Michael Taylor
Cameron Colby Thomson
Tom Veal
Sam Vilain
Guy A. Wadsworth
Sissy Willis
Rick Wilson

Honored Donors
That used pseudonyms[3]

"Backer Name"   "Rob"
"David"   "Sagavia"
"MegaZone"   "Scott K"
"Mister Generic"

---

3 For the most part, the names come directly from the Kickstarter web site and some are pseudonyms. A best effort was made by the author to establish who the contributors actually were, but some could not be identified, or preferred to go by their pseudonym (which, in this listing, appears within quotation marks). Of course, the author cannot guarantee that any of the contributer-provided names are not pseudonyms.

Appendix A:

A Guide To The Government Manned Space Program 1981 – 2013

There has been a great deal of confusion, ignorance and outright lies about what is going on in space policy since ... well ... forever. To try to remedy this, I decided that it would be useful to put together a little guide, so that people could understand the old plans versus the current ones, and have a better basis for determining whether or not there is any improvement. This is also intended to provide needed context for those who are unfamiliar with them.

Space Shuttle

What it was:

The Space Shuttle (also known as the National Space Transportation System, or NSTS) was the means by which NASA got its astronauts and cargo to and from low earth orbit (LEO) for the past three decades or so. It flew from the spring of 1981 through the summer of 2011. It was a specific vehicle design that had a payload bay that could carry tens of tons to low earth orbit and return somewhat less back to earth. It could carry up to seven astronauts, had the ability to stay in orbit for up to two weeks and could even act as a short-term space station.

It had a robotic arm to deploy and retrieve payloads as necessary and an airlock that allowed astronauts to perform spacewalks for satellite repair and other extravehicular activities (EVA). The orbiter part was reusable, the two first-stage solid-rocket boosters (SRBs) were retrieved and rebuilt after each flight, but the large central propellant tank was expended on every launch.

It was originally intended to perform all space transportation services for the nation. However, after the *Challenger* disaster in 1986—which caused a moratorium on flights for almost three years, during which NASA had no access to space, until the problems could be resolved—it was recognized that putting all our eggs in one "Shuttle" basket was an unrealistic and dangerous goal.

What it is not:

The "Space Shuttle" is not a generic term, like "kleenex," or "xerox," or "google," to describe any vehicle that takes NASA astronauts to space and back. There will never be a "replacement Space Shuttle," or "NASA's next Shuttle," because NASA will never again build and operate a single vehicle type to carry out a multitude of various mission requirements. In the future, numerous separate vehicle types and orbital facilities will replace all of the Shuttle's combined functions—with redundancy.

Vision for Space Exploration (VSE)

What it was:

VSE was the new policy that was declared by President George W. Bush on January 14th, 2004—not quite a year after the loss of the Space Shuttle *Columbia*—and it was a consequence of that disaster. Prior to that date, the official policy of the NASA human spaceflight program had been to complete the International Space Station (ISS), and then to utilize the ISS until a decision was made to change that policy. There had been no plans to send humans beyond LEO and, in fact, through much of the 1990s, NASA had been expressly forbidden by Congress to even contemplate such things, because Congress didn't want to be committed to such a seemingly expensive project.

That changed dramatically with the announcement of the VSE. NASA was now authorized to go beyond LEO; first to the moon, where they would learn how to live on another world and utilize its resources; and then on to destinations beyond the earth-moon system, including Mars and other places. The Space Shuttle would be retired in 2010 and NASA would use the funds thus freed up to develop a new system, called the "Crew Exploration Vehicle" (CEV), to get its astronauts into orbit and on to the moon and other places. It was to be operational in 2014, implying that there would be a "gap" of three years during which we would (necessarily) rely on the Russians for access to the ISS (as we did from 2003-2005, during the stand down of the Shuttle fleet caused by the *Columbia* disaster). By 2020, we would once again have NASA astronauts on the moon, this time to stay.

What it was not:

VSE was not Constellation (see below). It was not any particular implementation of the expressed goals—neither the CEV operational date of 2014, nor the return to the moon by 2020. And it was not a specific plan to use the existing work force, launch elements or facilities, nor was it required to do so. In fact, following the announcement in 2004, a commission was assembled headed by government and industry aerospace executive Edward ("Pete") Aldridge, to provide guidance on how to implement the VSE. But it did not call for preservation of either the NASA/contractor work force or facilities, but rather encouraged international and commercial participation.

Constellation

What it was:

Constellation was the transportation architecture chosen by NASA administrator Michael Griffin in late 2005 to implement the goals of the VSE. It consisted of: 1) The Ares I Crew Launch Vehicle (CLV); 2) the Orion spacecraft; 3) the heavy-lift Ares V also known as the Cargo Launch Vehicle, or CaLV; 4) an Earth Departure Stage (EDS) to propel the Orion from LEO to the moon; and 5) a lunar lander named "Altair" to get to and from the lunar surface in a fashion similar to the Apollo LEM. It is worth noting that in selecting this architecture, NASA essentially ignored many of the Aldridge recommendations.

The Ares I CLV utilized a solid-propellant first stage (a modified Shuttle SRB) and had a liquid-propellant upper stage. The Orion spacecraft (the new name for the CEV proposed from the VSE) went on top of it, and then on to the moon. The heavy-lift Ares V was based on elements of the Ares I. Its first stage combined SRBs with a Shuttle-derived propellant tank and engines.

Constellation was cancelled in early 2010. At that time, only the Ares I CLV and Orion were under active development. This was because funds weren't available for the other elements as the Shuttle had not yet been retired. Orion was furthest along in development, but Ares I was having schedule and technical issues, and it wasn't expected realistically to be ready prior to 2017, which added at least three years to "the gap" during which we would have to rely on the Russians for transportation to and from the ISS. *In addition, its budget was ballooning, and its schedule was slipping more than a year per year.*

What it was not:
Constellation was not the VSE, though many associated and even equated it with the VSE, and even human spaceflight itself. It was simply a particular and hyperexpensive transportation architecture to implement it. Other architectures could have (and I think should have) done as well or much better, in terms of cost, schedule and sustainability. (This includes, for example, the new approach proposed in the proposed Obama 2011 budget, to use existing rockets with new in-space technologies).

Constellation was not a "Space Shuttle replacement" (see Space Shuttle). It was both more and less than that. It replaced the Shuttle's capability to get crew to and from orbit, and the lofting of large payloads. And it was, at least in theory, an entire architecture to get humans all the way to the lunar surface and back, something that the Shuttle had never been able to do. But it didn't replace the Shuttle's other unique capabilities, such as large payload return and orbital research and operations. And the total cost for Constellation was projected to be much greater than thirty billion considering that thirty billion dollars was the price tag for just the Ares I rocket alone.

Constellation was not just the Ares launch systems and Orion spacecraft, but those are the elements that people are still fighting to preserve, because there are many jobs at stake in several states.

Constellation was cancelled in its entirety by the Obama administration in 2010. But the only practical effect was to cancel Ares I and Orion, because everything else was just fairy dust at the time.

Orion

What it is:

Orion is an Apollo-type spacecraft capable of carrying seven astronauts to LEO, and four to the moon. Originally it was the manned, Crew Exploration Vehicle (CEV) of the VSE policy. In 2005, the CEV was then named "Orion" as part of the Constellation program. After Constellation was canceled in early 2010, many in Congress protested, and in the fall of that year, a new authorization bill was passed that required NASA to develop, among other things, what Congress called a "Multi-Purpose Crew Vehicle" (MPCV). Its requirements, by no coincidence, were very similar to those for the Orion spacecraft of Constellation (though with some additions, such as the ability to do EVA) that essentially allowed work on it to continue uninterrupted. In the spring of 2011, MPCV was renamed back to "Orion." Basically Orion is a scaled-up Apollo capsule and Service Module that NASA claims is necessary for deep-space exploration with humans—though it is far too small to be psychologically practical for a multi-month mission. The current plan is for it to have a test flight to LEO on a United Launch Alliance (ULA) Delta IV rocket in 2014. But at the time of this book's publication (early summer of 2013), it is 5,000 pounds overweight, and the entry heat shield has recently suffered cracking under a stress test.

What it is not:

Orion is not a "Shuttle replacement." It is also not for simply getting crew to and from orbit, nor is it essential for deep-space exploration, though it may be useful for it, particularly for operations in cis-lunar space.

Space Launch System (SLS)

What it is:

SLS is a heavy-lift system, ultimately to match the performance of the Saturn V of the sixties, to launch cargo and humans in various mix-and-match configurations of common system elements. The Congress and particularly the Senate, wanted to resurrect the Ares V heavy-lift program from Constellation in some form, because it represented a significant part of the Shuttle work force in the states of those on the Senate space committee; most notably Bill Nelson of Florida, Kay Bailey Hutchison of Texas (who retired in 2012), Richard Shelby of Alabama and Orrin Hatch of Utah. So as part of the 2010 authorization bill, along with the Multi-Purpose Crew Vehicle (MPCV) that would later be renamed back to "Orion," they demanded that NASA build what they called the "Space Launch System" or SLS. From the legislative language, it was to have:

(A) The initial capability of the core elements, without an upper stage, of lifting payloads weighing between 70 tons and 100 tons into low-Earth orbit in preparation for transit for missions beyond low-Earth orbit.

(B) The capability to carry an integrated upper Earth departure stage bringing the total lift capability of the Space Launch System to 130 tons or more.

(C) The capability to lift the multipurpose crew vehicle.

(D) The capability to serve as a backup system for supplying and supporting ISS cargo requirements or crew delivery requirements not otherwise met by available commercial or partner-supplied vehicles.

But the more important requirements were these:

The Administrator shall ensure critical skills and capabilities are retained, modified, and developed, as appropriate, in areas related to solid and liquid engines, large diameter fuel tanks, rocket propulsion, and other ground test capabilities for an effective transition to the follow-on Space Launch System.
In developing the Space Launch System pursuant to section 302 and the multi-purpose crew vehicle pursuant to section 303, the Administrator shall, to the extent practicable utilize–

(1) existing contracts, investments, workforce, industrial base, and capabilities from the Space Shuttle and Orion and Ares 1 projects, including–

(A) space-suit development activities for application to, and coordinated development of, a multi-purpose crew vehicle suit and associated life-support requirements with potential development of standard NASA-certified suit and life support systems for use in alternative commercially-developed crew transportation systems; and

(B) Space Shuttle-derived components and Ares 1 components that use existing United States propulsion systems, including liquid fuel engines, external tank or tank-related capability, and solid-rocket motor engines; and

(2) associated testing facilities, either in being or under construction as of the date of enactment of this Act.

In other words, it was about preserving the existing work force and infrastructure, regardless of whether or not this made economic sense, or if it would result in a cost-effective system.

Appendix B provides a modified version of an extensive critique of the SLS that I wrote at the time it was authorized by Congress in the fall of 2010. As of mid-summer, 2013, it just completed Preliminary Design Review with a first flight planned in 2017. However, as I discuss in the book, it has a grim future, from a budget standpoint. It also has numerous technical issues, but none that can't be overcome with sufficient funding. The sequester of 2013 threatens it in the 2014 budget, and there is already talk among Senate appropriators of raiding the budget of the Commercial Crew program to pour it into the fiscal black hole that is SLS.

What it is not:

SLS is not a replacement for the Shuttle, and it is not necessary to send humans beyond low earth orbit. This can be accomplished much more cost effectively with multiple launches of smaller, much cheaper and existing rockets.

Commercial Space

What it is:

Commercial Space is launch providers who offer vehicles that weren't developed and aren't operated by NASA, and who can offer their services to other customers. Examples include United Launch

Alliance (ULA) with their Atlas V family developed by Lockheed Martin and Delta IV family developed by Boeing; Space Exploration Technologies (SpaceX) with their Falcon family of launchers; and the Orbital Sciences Corporation with their Antares and other smaller launchers. As already noted, ULA has many successful launches of multi-hundred-million-dollar satellites under its belt, and Boeing (which has a heritage of manned systems going back to Mercury, Gemini and Apollo) is developing an "Orion-lite" capsule under the Commercial Crew program that it calls "CST-100." In addition, Sierra Nevada Corporation is developing a small winged vehicle named "Dream Chaser" that will go up on an Atlas V, and return like the Shuttle and land on a runway.

SpaceX has successfully flown its launch vehicle, Falcon 9, several times, along with its "Dragon" crew/cargo module that has made multiple deliveries of cargo to and from the International Space Station. To carry passengers, the Dragon awaits only the implementation of the launch-abort system, expected to occur by 2015.

Having this multiplicity of providers gives us much more robust capability, in which the loss of a single vehicle type (e.g., Ares I) will not result in a multi-year stand down of the U.S. human spaceflight program, as has happened twice with the Shuttle over the years.

What it is not:
Commercial Space is not simply SpaceX. So people who want to equate it with SpaceX in order to declare the commercial space industry "unproven" don't know what they're talking about. And actually, all of the companies in the commercial space community comprise most of NASA's expertise in human spaceflight—including SpaceX—because they've been hiring NASA vets, including astronauts, like crazy.

Commercial Crew

What it is:

The Commercial Crew Program, or CCP, later transitioned to Commercial Crew Development (CCDev), and is now (Summer 2013) called Commercial Crew Integrated Capability (CCiCap). It is a program to provide NASA with multiple redundant providers to get its astronauts to and from the ISS. It is managed by the Commercial Crew and Cargo Program Office (C3PO—yes, really). It is intended to eliminate our dependence on the Russians for crew transfer and lifeboat services as soon as possible. It will also allow an increase in ISS crew complement by providing larger lifeboats. Current contenders in the program are: SpaceX with Falcon 9 and Dragon; Boeing with its CST capsule, which can fly on either Falcon 9 or ULA's Atlas V or Delta IV; and Sierra Nevada with its Dream Chaser launched on an Atlas V. All contracts are executed via Space Act Agreements with NASA, in a public/private partnership in which the companies provide their own resources, in addition to being reimbursed on a fixed-price basis for pre-agreed milestones. The program is on track as well as can be expected, given that Congress chronically underfunds it, with a first crew capability from SpaceX currently expected in 2015.

What it is not:

The Commercial Crew Program is not a "subsidy to send millionaire tourists to space," despite some of the nonsensical criticism of it by defenders of the status quo. While Boeing may not continue its efforts absent NASA funding, both SpaceX and Sierra Nevada seem determined to move forward to provide commercial passenger service to and from orbit, with their own funds if necessary. There is no doubt that NASA's contributions accelerate the program, because NASA has a need for it to be accelerated, but NASA needs the providers much more than they need NASA, at least in the case of SpaceX.

Appendix B

The "Senate" Launch System[1]

In late September of 2010, Congress passed a four-year authorization for NASA[2] which, among other things, stipulated that the agency must build a heavy-lift launch vehicle by 2016 based on existing (expensive) Space Shuttle components. But it did not authorize sufficient funds to accomplish this, at least if done in the manner that Congress had demanded. Specifically, Sections 302-304 of the authorization bill are particularly problematic and should be of great concern for both the taxpayer and those interested in sustainable progress in the nation's space endeavors. We'll start with Section 302: the Space Launch System:

SEC. 302. SPACE LAUNCH SYSTEM AS FOLLOW-ON LAUNCH VEHICLE TO THE SPACE SHUTTLE.

(a) United States Policy. — It is the policy of the United States that NASA develop a Space Launch System as a follow-on to the Space Shuttle that can access cis-lunar space and the regions of space beyond low-Earth orbit in order to enable the United States to participate in global efforts to access and develop this increasingly strategic region.

First, let's note that there will be no follow-on to the Space Shuttle if by that they mean a single vehicle that can perform

---

1 SLS is actually the "Space Launch System." However, due to the U.S. Senate's undesirable overinvolvement with its specifications, many have pejoratively referred to it as "the 'Senate' Launch System."

2 "National Aeronautics and Space Administration Authorization Act of 2010", http://www.commerce.senate.gov/public/?a=Files.Serve&File_id=20a7a8bd-50f4-4474-bf1d-f0a6a8824b01

all the functions that the Shuttle did. The Shuttle could carry large payloads to orbit and also return them to earth. It could act as a temporary space station for up to two weeks, and operate as a base for orbital repairs. There will never again be a Space Shuttle, because a single vehicle design that has to serve such a wide variety of conflicting requirements is a fragile system and ultimately becomes unaffordable. What one suspects they meant is the next system that NASA will develop, own, and operate as its sole means of getting to orbit.

Instead of a single-vehicle approach, we should be seeking a variety of vehicles that can replicate and extend the individual functions of what the Shuttle did in a cost-effective and redundant manner. However, as we'll see below, this apparently was not their goal and they are insisting once more, in defiance of the lessons from the Shuttle, on a single vehicle type to serve a variety of functions.

    (b) Initiation of Development.—

        (1) In general.—The Administrator shall, as soon as practicable after the date of the enactment of this Act, initiate development of a Space Launch System meeting the minimum capabilities requirements specified in subsection (c).

        (2) Modification of current contracts.—In order to limit NASA's termination liability costs and support critical capabilities, the Administrator shall, to the extent practicable, extend or modify existing vehicle development and associated contracts necessary to meet the requirements in paragraph (1), including contracts for ground testing of solid-rocket motors, if necessary, to ensure their availability for development of the Space Launch System.

Note the assumptions here: 1) the taxpayer will be best served by limiting termination liability costs (the costs of shutting down a contract prematurely before completion); and 2) that solid motors are "critical capabilities" to building the Space Launch System.

But limiting termination liability is a false economy if extending the contracts into the indefinite future increases NASA costs beyond what they might have been if they started with a different contract and concept. This—like the complaint that in canceling Constellation we will have "wasted" the ten billion dollars spent on it—is an example of the sunk-cost fallacy. Note also that, even if the argument wasn't fallacious, the same people who make such complaints about moneys invested apparently had no problem with the associated plan to abandon, if not actually deorbit, the International Space Station (ISS) in 2016—an "investment" into which the nation has sunk over one *hundred* billion dollars ($100,000,000,000).

In the face of numerous studies indicating otherwise, the Congress is assuming that the lowest-cost approach for the future is to continue the high-cost approach in which we've been engaged for the past half century. Note that SLS utilizes solid rocket boosters (SRB) similar to those of the Space Shuttle. But there are many heavy-lift rocket approaches that do not employ solid-rocket motors. It appears that the only thing for which SRBs are "critical" is the maintenance of a jobs base in the state of Utah.

Note also that, even though there is no defined mission for the system, the Congress knows exactly what the payload of the SLS should be. Well, sort of…

> (c) Minimum Capability Requirements.—
>
> > (1) In general.—The Space Launch System developed pursuant to subsection (b) shall be designed to have, at a minimum, the following:
> >
> > > (A) The initial capability of the core elements, without an upper stage, of lifting payloads weighing between 70 tons and 100 tons into low-Earth orbit in preparation for transit for missions beyond low-Earth orbit.
> > >
> > > (B) The capability to carry an integrated upper Earth departure stage bringing the total lift capability of the Space Launch System to 130 tons or more.
> > >
> > > (C) The capability to lift the multipurpose crew vehicle.

OK, so they want the "initial" capability to be at least 70 tons (the "ton" here is 2,000 pounds mass, not the metric ton, which is a thousand kilograms, or over 2,200 pounds mass), and up to 100 tons. That is, between 140,000 and 200,000 pounds of payload to low earth orbit (LEO). But then they say that they want the capability (though not specified as initial, so perhaps it can grow to this over time) to deliver even more—260,000 pounds of payload to...somewhere. This second capability is confusing. The bill doesn't have a definition for "total lift capability," so it's not clear what this means.

Note that they wanted to launch the multipurpose crew vehicle (MPCV—what used to be called Orion in Constellation, and is once again called Orion) with the SLS. Since Orion weighs far less than seventy tons, if the destination is to LEO, the lift effort will be trivial. But presumably, they also want Orion to be launched on top of the "Earth departure stage" to send it to locations beyond LEO. Finally, it should have...

> (D) The capability to serve as a backup system for sup-
> plying and supporting ISS cargo requirements or
> crew delivery requirements not otherwise met by
> available commercial or partner-supplied vehicles.

In other words, they wanted to use a heavy-lift (minimum 70-ton) vehicle to service the ISS. For an ISS mission, it would be unlikely for the MPCV/Orion to weigh more than thirty tons, even with its launch escape system, so the launch vehicle will be vastly oversized for this mission. Every credible cost analysis indicates that this vehicle will cost well over one billion dollars per flight, and perhaps two or three billion, when taking into account amortization of development and fixed annual costs. If the MPCV can carry six persons per mission, that comes out to a cost of almost two-hundred million per person, and that doesn't even count the cost of the MPCV (which is still unknown, and will depend a lot on whether or not it is reusable). But even ignoring this cost, which will surely be hundreds of millions (again, factoring in amortization), that is a ticket price three times that being charged by the Russians in their latest contract. And it's ten times what Space Exploration Technologies has quoted for a Falcon/Dragon flight (twenty million per ticket for a crew of seven). Now, they may argue that they can reduce the crew costs by booking some of the flight cost to cargo with the additional capacity, but this goes against the recommendation of the Columbia Accident Investigation Board (CAIB) that NASA never again mix crew and cargo on a flight.

Does any of this make sense?

And in this next requirement, they remain determined to repeat the mistakes of the Shuttle:

> (2) Flexibility.—The Space Launch System shall be designed from inception as a fully-integrated vehicle capable of carrying a total payload of 130 tons or more into low-Earth orbit in preparation for transit for missions beyond low-Earth orbit. The Space Launch System shall, to the extent practicable, incorporate capabilities for evolutionary growth to carry heavier payloads. Developmental work and testing of the core elements and the upper stage should proceed in parallel subject to appropriations. Priority should be placed on the core elements with the goal for operational capability for the core elements not later than December 31, 2016.

"Flexibility" was one of the things that made Shuttle so expensive, and it will be easy to make this new vehicle unaffordable in exactly the same way. Now, a minimal escape trajectory from earth orbit, say to the moon, requires a velocity change of about 3,500 meters per second. If we assume a stage fraction of 90% (that is, the propellant will be ninety percent of the total stage weight) and a specific impulse (a factor that indicates the fuel economy of a rocket engine) of 480 seconds for LOX/hydrogen propulsion, we could toss about fifty tons beyond low earth orbit and to the moon. Obviously, the amount of payload would be somewhat less to leave the region of the earth-moon system to go to an asteroid or Mars. Whether this is enough payload mass for an effective mission, let alone a cost-effective one, is far beyond the scope of this analysis, but it at least provides some basis for the payload number. In reality, multiple launches will probably be required to get a reasonably sized mission, negating the supposed advantage of building such a large (and costly, both to develop and operate) launch vehicle, which was supposed to avoid orbital mating and assembly.

But it's the next requirement that's the real one, as far as Congress is concerned:

> (3) Transition needs.—The Administrator shall ensure critical skills and capabilities are retained, modified, and developed, as appropriate, in areas related to solid and liquid engines, large diameter fuel tanks, rocket propulsion, and other ground test capabilities for an effective transition to the follow-on Space Launch System.
>
> (4) The capacity for efficient and timely evolution, including the incorporation of new technologies, competition of sub-elements, and commercial operations.

All of this is code for "preserve the Shuttle infrastructure and all the jobs associated with it." And the notion that this would ever be amenable to commercial operations, particularly in light of the fierce competition it will have from true cost-effective commercial operators, domestic and foreign, is ludicrous.

The next section, while not dealing with the SLS itself, is also problematic:

SEC. 303. MULTI-PURPOSE CREW VEHICLE.

(a) Initiation of Development.—

> (1) In general.—The Administrator shall continue the development of a multi-purpose crew vehicle to be available as soon as practicable, and no later than for use

with the Space Launch System. The vehicle shall continue to advance development of the human safety features, designs, and systems in the Orion project.

(2) Goal for operational capability. — It shall be the goal to achieve full operational capability for the transportation vehicle developed pursuant to this subsection by not later than December 31, 2016. For purposes of meeting such goal, the Administrator may undertake a test of the transportation vehicle at the ISS before that date.

(b) Minimum Capability Requirements.–The multi-purpose crew vehicle developed pursuant to subsection (a) shall be designed to have, at a minimum, the following:

(1) The capability to serve as the primary crew vehicle for missions beyond low-Earth orbit.

(2) The capability to conduct regular in-space operations, such as rendezvous, docking, and extra-vehicular activities, in conjunction with payloads delivered by the Space Launch System developed pursuant to section 302, or other vehicles, in preparation for missions beyond low-Earth orbit or servicing of assets described in section 804, or other assets in cis-lunar space.

(3) The capability to provide an alternative means of delivery of crew and cargo to the ISS, in the event other vehicles, whether commercial vehicles or partner-supplied vehicles, are unable to perform that function.

(4) The capacity for efficient and timely evolution, including the incorporation of new technologies, competition of sub-elements, and commercial operations.

Basically, the Congress wants NASA to continue the Orion crew module, originally designed as a lunar orbit vehicle under a different name, and once again, they complicate the design by imposing additional requirements on it, all while providing less money than Orion was going to cost. They demand that it be capable of competing with commercial vehicles that can perform the ISS support missions. This is in violation of NASA's charter, the Space Act, that states "The Congress declares that the general welfare of the United States requires that the National

Aeronautics and Space Administration (as established by title II
of this Act) seek and encourage, to the maximum extent possible,
the fullest commercial use of space." One does not "encourage"
even to a minimum extent "the fullest commercial use of space"
by funding and subsidizing competition to commercial vehicles
with the taxpayers' money, even for government uses. It is an even
more blatant violation of the 1998 Commercial Space Act, whose
specific purpose was the promotion of commercial space activities,
particularly with regard to the International Space Station. The act
is quite explicit in this regard:

> The Congress declares that a priority goal of constructing the
> International Space Station is the economic development of
> Earth orbital space. The Congress further declares that free and
> competitive markets create the most efficient conditions for
> promoting economic development, and should therefore govern
> the economic development of Earth orbital space. The Congress
> further declares that the use of free market principles in operating,
> servicing, allocating the use of, and adding capabilities to the
> Space Station, and the resulting *fullest possible engagement of
> commercial providers* and participation of commercial users, will
> reduce Space Station operational costs for all partners and the
> Federal Government's share of the United States burden to fund
> operations. [Emphasis added]

There is nothing that I'm aware of in the 2010 authorization
declaring this law null and void so, as it stands, the two laws are
in conflict. Congress and the Clinton White House got it exactly
right in 1998, and a later Congress got it tragically wrong in 2010.
At the time of cancellation, Orion was projected to cost many
billions more to develop, for a mission with a more limited scope
(for instance, it had no extravehicular capabilities)—to get crew
to and from the moon. If it is redirected to doing additional LEO
missions as its new name—multi-purpose crew vehicle (MPCV)
would imply, its costs can only increase. If its lunar capabilities
are removed in an attempt to rein in the program costs, then it

is nothing except unfair competition for the commercial vehicles that are currently under development by Boeing, Sierra Nevada Corporation, Space Exploration Technologies, Orbital Sciences and others, at a huge cost to the taxpayer. This is not only not in keeping with the promotion of commercial enterprise—it is the very *antithesis* of it.

But all of this language to date is really an excuse for what it really desired, which is stated explicitly in the next section, 304. Absent this section, Congressional interest in the SLS and MPCV would pretty much evaporate, because the real goal is simply to maintain the Shuttle Industrial Complex, the aluminum/graphite-epoxy triangle of Congress, NASA and industry that has kept the taxpayers' money flowing in certain directions for decades, regardless of results:

SEC. 304. UTILIZATION OF EXISTING WORKFORCE AND ASSETS IN DEVELOPMENT OF SPACE LAUNCH SYSTEM AND MULTI-PURPOSE CREW VEHICLE.

(a) In General.–In developing the Space Launch System pursuant to section 302 and the multi-purpose crew vehicle pursuant to section 303, the Administrator shall, to the extent practicable utilize–

    (1) existing contracts, investments, workforce, industrial base, and capabilities from the Space Shuttle and Orion and Ares 1 projects, including–

        (A) space-suit development activities for application to, and coordinated development of, a multi-purpose crew vehicle suit and associated life-support requirements with potential development of standard NASA-certified suit and life support systems for use in alternative commercially-developed crew transportation systems; and

        (B) Space Shuttle-derived components and Ares 1 components that use existing United States propulsion systems, including liquid fuel engines, external tank or tank-related capability, and solid-rocket motor engines; and

    (2) associated testing facilities, either in being or under construction as of the date of enactment of this Act.

(b) Discharge of Requirements.–In meeting the requirements of subsection (a), the Administrator–

    (1) shall, to the extent practicable, utilize ground-based manufacturing capability, ground testing activities, launch and operations infrastructure, and workforce expertise;

    (2) shall, to the extent practicable, minimize the modification and development of ground infrastructure and maximize the utilization of existing software, vehicle, and mission operations processes;

    (3) shall complete construction and activation of the A-3 test stand with a completion goal of September 30, 2013;

    (4) may procure, develop, and flight test applicable components; and

    (5) shall take appropriate actions to ensure timely and cost-effective development of the Space Launch System and the multi-purpose crew vehicle, including the use of a procurement approach that incorporates adequate and effective oversight, the facilitation of contractor efficiencies, and the stream-lining of contract and procurement requirements.

I would note that (1) through (4) strongly conflict with (5). It is simply not possible to engage in a "timely and cost-effective development" of SLS and Orion using legacy infrastructure and contracts. Constellation was overrunning by billions, and slipping more than a year per year in schedule. Insistence on simply extending existing (and in many cases sole-source, without competition) contracts will fail just as surely. This was pointed out by the General Accountability Office (GAO), the Aerospace Corporation which is a Federally Funded Research and Development Center (FFRDC) for the United States Air Force, and the Augustine Committee that was convened by the Obama administration in 2009 to address the budget and schedule issues with the program. These issues won't go away just because Congress passes a law dictating that they must, any more than Congress can mandate a different value of gravity.

All of these requirements, from specifying vehicle size and MPCV functions, to how they should do it, and with what infrastructure and contractors, are far below the pay grade or competence of congresspeople and their staff. They may have many skills and talents, but they are not rocket scientists. Such requirements should be the result of engineering trade studies performed by NASA with the aid and input of the commercial contractor community. But, because our actual progress in space is not nationally important, what should be technical decisions have become political ones.

To summarize, anyone interested in actually developing space should oppose these aspects of the 2010 NASA authorization bill. In particular, they should oppose attempted congressional mandates to impose a design solution on NASA for either a heavy-lift vehicle or a traditionally-procured crew vehicle, particularly given that: 1) the requirements for such systems are poorly defined and justified; and 2) there are no payloads planned or funded for them. Such attempts appear to be a means of preserving an infrastructure in certain congressional districts and states, with high legacy costs, and associated legacy jobs. Numerous analyses, by the Aerospace Corporation, the General Accountability Office, and the Congressional Budget Office, have indicated that the type of system demanded by Congress in 2010 is unaffordable and unsustainable in today's austere budget environment. If there is a need for these systems, NASA should be allowed to do the trade studies that it has initiated, and determine the best means to go forward in a competitive manner to get the best deal for the taxpayer, without presumed design mandates driven by politics rather than mission or cost effectiveness.

Acronyms

AIAA	American Institute of Aeronautics and Astronautics
AOA	Abort Once Around
ASAP	Aerospace Safety Advisory Panel
ASTP	Apollo-Soyuz Test Project
ATO	Abort To Orbit
BASE	Buildings Antennas Spans and Earth
C3PO	Commercial Crew and Cargo Program Office
CAIB	Columbia Accident Investigation Board
CaLV	Cargo Launch Vehicle
CASIS	Center for the Advancement of Science in Space
CCDev	Commercial Crew Development
CCiCAP	Commercial Crew Integrated Capability
CCP	Commercial Crew Program
CEV	Crew Exploration Vehicle
CLV	Crew Launch Vehicle
CM	Command Module
COMSTAC	Commercial Space Transportation Advisory Committee
COTS	Commercial Orbital Transportation Services
CSF	Commercial Spaceflight Federation
CSM	Command and Service Modules
CST	Crew Space Transportation
DOT	Department of Transportation
EDS	Earth Departure Stage
EECOM	Electrical, Environmental and COMmunication systems
EELV	Evolved Expendable Launch Vehicle
ESAS	Exploration Systems Architecture Study
ET	External Tank
ET SEP	External Tank Separation
EVA	Extravehicular Activity
FAA	Federal Aviation Administration
FAA-AST	FAA-Office of Commercial Space Transportation [sic]
FAR	Federal Acquisition Regulations
FDA	Food and Drug Administration
FFRDC	Federally Funded Research and Development Center
HST	Hubble Space Telescope
ICA	Independent Cost Assessment
ICBM	Intercontinental Ballistic Missile
IEEE	Institute of Electrical and Electronics Engineers
INKSNA	Iran/North-Korea/Syria Non-Proliferation Act
IOC	Initial Operational Capability

IRBM	Intermediate-Range Ballistic Missile
ISS	International Space Station
ITAR	International Traffic in Arms Regulations
KSC	Kennedy Space Center
LAS	Launch-Abort System
LEM	Lunar Excursion Module
LEO	Low Earth Orbit
LLRV	Lunar Lander Research Vehicle
LOC	Loss Of Crew
LOX	Liquid Oxygen
MECO	Main Engine Cut Off
MER	Mars Exploration Rover
MPCV	Multi-Purpose Crew Vehicle
MRSA	Methicillin-Resistant Staphylococcus Aureus
MSFC	Marshall Space Flight Center
NACA	National Advisory Committee on Aeronautics
NASA	National Aeronautics and Space Administration
NASTAR	National Aerospace Training & Research Center
NOTAM	NOtice To AirMen
NSTS	National Space Transportation System
OCST	Office of Commercial Space Transportation
OMB	Office of Management and Budget
OMS	Orbital Maneuvering System
OST	Outer Space Treaty
PRA	Probabilistic Risk Assessment
RCS	Reaction Control System
ROI	Return On Investment
RTLS	Return To Launch Site
S&MA	Safety and Mission Assurance
SAA	Space Act Agreement
SEI	Space Exploration Initiative
SLS	Space Launch System
SRB	Solid-Rocket Booster
SS2	SpaceShipTwo
SSBPI	Spaceflight Safety Best Practices Institute
SSME	Space Shuttle Main Engine
STScI	Space Telescope Science Institute
TAL	Trans-Atlantic Abort
UKSA	United Kingdom Space Agency
ULA	United Launch Alliance
USN	United States Navy
USOS	United States Orbital Segment (Part of the ISS)
USSG	United States Space Guard
VSE	Vision for Space Exploration

Illustration Credits

Chapter 1

Pg. 2   Ferdinand Magellan, 16th or 17th century anonymous portrait.
The Mariner's Museum Collection, Newport News, VA. In the public domain.

Pg. 2   Ernest Henry Shackleton, R.N.R, Sub-Lieut, aged 27.
In the public domain

Pg. 4   Jesse William Lazear
In the public domain

Pg. 5   Joseph Kittinger's jump 1960
Credit: US Air Force/Volkmar Wentzel.

Pg. 6   Portrait of Colonel John Paul Stapp
Credit: USAF photo. Courtesy of New Mexico Museum of Space History

Pg. 6   John Paul Stapp undergoes deceleration in rocket sled test, December 10 1954
Credit: USAF photo. Courtesy of New Mexico Museum of Space History

Pg. 10   NASA STS-130 Astronaut Nicholas Patrick during ISS EVA, February 17, 2010
Credit: NASA

Pg. 10   Apollo 17 Astronaut Harrison H. Schmitt with the Lunar Roving Vehicle (LRV)
at the Taurus-Littrow landing site, December 12, 1972
Credit: Eugene A. Cernan/NASA.

Pg. 10   Curiosity's Mast Camera panorama looking east from the Glenelg/Rocknest site
in Gale Crater, October/November 2012
Credit: NASA/JPL-Caltech/Malin Space Science Systems

Chapter 2

Pg. 11   Trevithick Locomotive, 1801
In the public domain

Pg. 13   Orville piloting the Wright Flyer III at Ft. Myer, VA September 1908
Credit: Unknown US Army photographer.

Pg. 13   Blériot XI (1999 replica by Hans Furrer, original motor and wheels, HB-RCV) at
"Oldtimer Fliegertreffen Hahnweide 2011" September 2011
Credit: Julian Herzog.

Pg. 14   The JN-4 "Jennie"
In the public domain

Chapter 4

Pg. 41   ULA Atlas V on launch pad at Cape Canaveral, June 17, 2009
          Credit: NASA/Jack Pfaller

Pg. 41   ULA Atlas V(401) lift-off from Cape Canaveral, June 18, 2009
          Credit: NASA

Pg. 41   SpaceX Falcon 9 on launch pad at Cape Canaveral
          Credit:

Pg. 41   SpaceX Falcon 9 lift-off from Cape Canaveral
          Credit: NASA

Chapter 5

Pg. 45   Ares I Crew Launch Vehicle (CLV)
          Credit: MSFC/NASA

Pg. 47   Fig. 1: Establishing Crew Safety Goals - the value of an escape system
          Credit: NASA/http://www.scribd.com/doc/18112812/NASA-Ares-I-V-Launch-
          Vehicle-July-2009-Status-Report

Pg. 48   Fig. 2: Loss of Crew Contributors for Earth Orbit Rendezvous-Lunar Orbit
          Rendezvous 1.5 Launch Mission
          Credit: NASA/http://www.nasa.gov/pdf/140649main_ESAS_full.pdf Page 568

Pg. 52   Orion Launch-Abort System drawing
          Credit: Based on graphic from Astronautix www.astronautix.com

Pg. 62   Proposed Orion Capsule
          Credit: NASA

Pg. 65   The Space Launch System (SLS)
          Credit: NASA

Chapter 6

Pg. 71   The International Space Station as seen by STS-129, November 2009
          Credit: NASA/Crew of STS-129

Pg. 72   Soyuz TMA-9 Launch - September 18, 2005
          Credit: NASA/Bill Ingalls.

Pg. 72   Soyuz TMA-7 Spacecraft
          Credit: NASA

Pg. 73   SpaceX Dragon in orbit
          Credit: NASA

Chapter 7

Pg. 85   The Hubble Space Telescope (HST) in orbit
          Credit: NASA/Space Science Telescope Institute (STSci)

Pg. 90   Hubble Space Telescope images (as arranged on page 90):

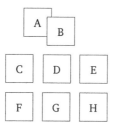

A: Palomar image of M33
      Credit: Ground-based image courtesy of Palomar Observatory, Caltech, and the
      STScI Digitized Sky Survey

B: HST image of NGC 604 in M33
      Credit: 8/7/1996 Hui Yang (University of Illinois) and NASA

C: HST image of the Crab Nebula
      Credit: NASA, ESA, J. Hester and A. Loll (Arizona State University)

D: HST image of the Cat's Eye Nebula
      Credit: NASA, ESA, HEIC, and The Hubble Heritage Team (STScI/AURA)
      Acknowledgment: R. Corradi (Isaac Newton Group of Telescopes, Spain) and
      Z. Tsvetanov (NASA)

E: A Rose of Galaxies
      Credit: NASA, ESA, and the Hubble Heritage Team (STScI/AURA)

F: Mystic Mountain
      Credit: NASA, ESA, M. Livio and the Hubble 20th Anniversary Team (STScI)

G: Butterfly Nebula (NGC 6302)
      Credit: NASA, ESA, and the Hubble SM4 ERO Team

H: Eskimo Nebula (NGC 2392)
      Credit: NASA, Andrew Fruchter and the ERO Team [Sylvia Baggett (STScI),
      Richard Hook (ST-ECF), Zoltan Levay (STScI)]

Notes: AURA   Association of Universities for Research in Astronomy
          HEIC   Hubble European Space Agency Information Centre
          SM4 ERO   Servicing Mission 4 Early Release Observations
          ST-ECF   Space Telescope European Coordinating Facility
          STScI   Space Telescope Science Institute

Chapter 8

Pg. 102 Actin and vinculin in adherent monocyte, in an ISS lab, March 26, 2013
           Credit: ESA/MIA G.Pani

Pg. 107 SpaceX Dragon spacecraft at SpaceX facility, June 14, 2012
           Credit: Bill Simon

Pg. 111 Golden Spike lunar lander concept
           Credit: Golden Spike Company

Pg. 112 Inspiration Mars capsule and habitat module (artist's concept), April 20, 2013
           Credit: Inspiration Mars Foundation

Chapter 9

Pg. 128 SpaceShipTwo suspended under WhiteKnightTwo, December 7, 2009
           Credit: Mark Greenberg/Virgin Galactic.

Pg. 131 Armadillo four-engine rocket concept
           Credit: Armadillo Aerospace

Pg. 134 The wingsuit: Scott and Mike over the coast, November 19, 2006
           Credit: Matt Hoover

Pg. 167 Opening the Space Frontier, The Next Giant Step
           Credit: NASA

———————————————

Pg. 215 Author's photo
           Credit: Bill Simon

Cover

Earth   Credit: NASA
Moon   Credit: Bill Simon, December 27, 2012
Mars   Credit: NASA, ESA, and The Hubble Heritage Team (STScI/AURA)
Design: Bill Simon

Index

A

Aaron, John 27
abandon (the ISS) 73, 82, 83
abort 23, 24, 28, 33, 34, 46, 50-53, 55,
    106, 107, 114, 126, 129, 146-
    151, 179. *See also* Launch-Abort
    System
Abort Once Around 34
Abort To Orbit 34
active dampers 51
acceleration xxiii, 6, 24, 146, 147
accident xv, 13, 16, 17, 24, 47, 124, 133,
    143, 144, 146
actuators xxii, xxiii,
Adams, Michael 19
Admiral Gehman *See* Gehman, Harold
Aedes aegypti mosquito 4
aerialists 14, 15
aerodynamic xxii, 30, 37
Aerospace Corporation 51, 190, 191
Aerospace Safety Advisory Panel 95, 116,
    121
aerosurface 126
Adams, Michael 19
Afghanistan 108
AIAA 139
airbags (landing device) 67
Air Commerce Act of 1926 16,
aircraft xvii, xxi, xxiii, 4, 6, 13, 14, 16, 19,
    20, 23, 24, 39, 43, 76, 78, 79, 82,
    100, 124, 126, 128-130, 134, 158
Air Force 4, 6, 13, 19, 21, 22, 83, 98, 100,
    118, 161, 190
air launched 129
airline 16, 39, 40, 150
airliner 17, 35
airlock 31, 171
airship 14
airworthiness 16
Alps 13
Altair 173
altitude xxi, 4-6, 14, 20, 23, 35, 75, 79,
    126, 128, 135-137
ambulance 74, 75, 76

American Institute of Aeronautics and
    Astronautics 139
American Society of Mechanical
    Engineers 12
Amundsen, Roald 14
Amundsen-Scott 75, 77, 102
Analytical Hierarchy Process 68
Anders, Bill 26
Anderson, Michael xxiv,
Angel, Eddie 15
Antarctica 75, 76
Antarctic Station 77
Antares 179
AOA 34
aplastic anemia 4
Apollo viii, xvii, xviii, 23-33, 38, 55, 91,
    95, 96, 106, 109, 111-113, 115,
    136, 154, 156, 157, 160, 162, 163,
    174, 175, 179
Apollo-1 24, 95, 106
Apollo-8 25, 26, 29, 30, 108, 112, 115
Apollo -12 27
Apollo-13 29, 30
Apollo-Soyuz Test Project 31
Arctic xvii, 10, 14
Ares I 45, 46, 49, 51, 54-56, 62, 98, 108,
    118, 119, 160, 173-175, 179
Ares I-X 118, 119
Ares V 173, 174, 176; Heavy-lift 46, 62
Ares program 118
Ares/Orion 48, 110
Arlington 162
armada 109, 112
Armadillo Aerospace 125, 130, 131, 136,
    146, 147
Armstrong, Neil 20
Army Air Corps 6
ASAP 95, 116, 118-121
ASME 12
ascent 5, 22, 25, 29, 33, 48-51, 80, 92,
    136,
asphyxiated 24,
Associated Press 93, 115
assumed risk 145,
AST 132, 135, 136, 145, 159
asteroid 99, 117, 186

About The Author

Rand Simberg is a recovering aerospace engineer with over a third of a century of experience in the space industry. Early in his career, he accumulated over a decade of experience in engineering and management at the Aerospace Corporation in El Segundo, California and Rockwell International in Downey, California. Since leaving Rockwell in 1993, he has been a consultant in space technology and business development as well

Rand Simberg

as a technology entrepreneur. He also advises on regulatory and market issues pertaining to commercial and personal spaceflight.

Mr. Simberg holds multiple engineering degrees from the University of Michigan, Ann Arbor and a Masters degree in Technical Management from West Coast University in Los Angeles. He is an adjunct scholar with the Competitive Enterprise Institute, and has written many pieces for *Popular Mechanics, Fox News, America Online, PJMedia, National Review, Reason* magazine, *The Weekly Standard, The Washington Times*, and *TCSDaily*, among others. He has also written extensive essays on space policy and technology for the quarterly journal, *The New Atlantis*.

This is his first, but hopefully not last, book.

Revisions

Revision A   January 2014

1. Title Page:
   - Is: "First Edition: October 2013 Revision A: January 2014"
   - Was: "First Edition Unrevised: October 2013"

2. Table of Contents: Added "Revisions page 217"

3. Page 126:
   - Is: "total loss of thrust"
   - Was: "premature engine shutdown"
   - Reason: Original wording misleadingly implied that system doesn't have engine-out capability on takeoff, though it has multiple engines.

4. Page 203:
   - Is: "Bigelow Aerospace"
   - Was: "Bigelow Aerosapce"

CPSIA information can be obtained at www.ICGtesting.com
Printed in the USA
BVOW11s2319140214

344669BV00003BA/11/P